# Nomenclature

Popular names of trees vary widely; this makes them unreliable, and accuracy is best served by employing the scientific (Latin) name, the structure of which is governed by strict international rules. All trees have a double name: the first element is that of the genus, and the second the name of the species within the genus. For example, the birches are known as the *Betula* genus, and the River birch is the *nigra* species, and so the River birch is known as *Betula nigra*. Genera with noticeable similarities are grouped into families, and *Betula* belongs in the Betulaceae.

Sometimes further definition is required beyond the species. A variety (indicated by "var." after the specific name) is a tree that has developed slightly different characteristics from its species (or "type") due to regional ecological factors. A cultivar (cv.) or clone is propagated vegetatively from a cutting from a tree that showed some minor genetic change. It is given a new name, which is printed in inverted commas. A hybrid (denoted by "x" between the generic and specific names) is a natural cross between two species. Where the parentage of a hybrid is in doubt, as with many *Prunus* hybrids, a popular name replaces the specific name.

### Acknowledgements

The author and publishers would like to thank the following people and organizations for their help:
**Alf Westall and Malcolm Scott,** Bedgebury National Pinetum, Kent; **Alan Mitchell,** The Forestry Commission, Alice Holt Lodge, Farnham, Surrey; **Jim Keesing and Charles Erskine,** The Royal Botanic Gardens, Kew; **Ivan Hicks,** West Dean Estates, Chichester, Sussex.

*Artwork*
**Olivia Beasley:** 24/25, 30/31, 34/51, 54/63; **John Davis:** 83/90, 107/110, 115/117, 119, 140/155; **John Michael Davis:** 23, 28/29, 32/33, 52/53, 64/79, 132/139, 156/181; **Annabel Milne and Peter Stebbing:** 5/8, 11/22, 26/27, 80/82, 182/211; **David Moore:** 1, 91/104, 120/131; **Paul Wrigley:** 105/106, 111/114, 118

**Editors** David Arnold and Ken Hewis
**Designer** Jacqueline Moore
**Executive Editor** Susannah Read **Art Editor** Douglas Wilson
**Production** Julian Deeming

Edited and designed by
Mitchell Beazley Publishers
87–89 Shaftesbury Avenue, London W1V 7AD
© Mitchell Beazley Publishers 1981
Printed in Hong Kong by Mandarin Offset International Ltd.
A Fireside Book
Published by Simon and Schuster
A Division of Gulf & Western Corporation
Simon and Schuster Building
Rockefeller Center
1230 Avenue of the Americas
New York, New York 10020

Library of Congress
Cataloging in Publication Data
Rushforth, Keith.
The pocket guide to trees.
1. Trees – Identification.
2. Trees – Europe –
Identification.

3. Trees – United States –
Identification.
I. Title.
QK477.2.I4R87
582.160909'813 80-5505
ISBN 0-671-25514-2

# Contents

# Introduction

This book is a field guide that will enable the reader to identify almost any tree he or she encounters. North America boasts over 800 native species, besides which many introduced species are familiar. In this book we have been able to illustrate over 400 species as well as mention other species, varieties and cultivars. This means that all commonly encountered trees, with the exception of those rarely found outside arboreta, should be readily identifiable.

There is a generally accepted order in which tree families should be placed, based upon their presumed evolutionary sequence from the more primitive to the advanced groups. Occasionally I have deviated from this in order to place side by side species which bear a strong resemblance to each other so that they may be more readily identified. Within families the genera have been ordered purely for the sake of convenience, and their sequence does not imply any botanical significance.

A tree is usually defined as a woody perennial plant growing on a single stem to a height of 6 m or more, whereas a shrub does not attain this height and has a stem divided near the ground. But of course plants do not fall obligingly into these man-made categories all the time, and some of them occur as either a tree or a shrub. The rule I have followed is that if a plant occurs as a tree with more than a negligible frequency, then it should be included. A plant included on these grounds, but often excluded from tree books, is Juniper. I have also tried to pay particular attention to the appearance of deciduous trees in winter, a phase of their existence all too often ignored in books.

Points about size can only be made in general terms; this particularly concerns the bole, which increases in diameter every year—a distinctive characteristic of the growth of trees. The girth of the trunk 1.5 m above the ground is a reasonable method of estimating the age of a tree, and the rate of growth shows a remarkable degree of consistency among the species. The average increase in girth of a tree growing in the open is 2.5 cm a year; thus, if it has a girth of about 2.5 m, it is about 100 years old. The growth rate of long-lived trees soon slows down; Ancient pines in California are over 4,000 years old. All measurements given in this book are maxima.

# How to use this book

The annotated illustrations, text and symbols, which are explained below, give in concise form all the information necessary to identify a tree, but you do not need laboriously to thumb your way through the book looking for the right species. The first step is to read the introductory pages, where the major differences between groups of trees are explained, and guidance is given on which parts of the tree to examine to narrow the choice. Once in the field you can use the four keys to identify the genus; from there you can proceed quickly to the species in question. Before you try to identify an unknown tree it is a good idea to practice using the keys by working backwards from one that is familiar.

More general matter will be found in the introduction to the broadleaf trees, p80.

## Symbols

| | | | |
|---|---|---|---|
| **R** | Rare, usually found only in collections | (symbol) | Needles in fascicles of 5 |
| (symbol) | Deciduous, i.e. leaves shed in autumn | (symbol) | Leaves in horizontal sprays along shoot |
| (symbol) | Evergreen, leaves retained into winter | (symbol) | Found in deciduous woods |
| (symbol) | Buds and leaves alternate or spiral | (symbol) | Found in evergreen woods |
| (symbol) | Buds and leaves in opposite pairs | (symbol) | Found in mixed woods |
| (symbol) | Leaves awl- or scale-like | (symbol) | Common in streets, parks and gardens |
| (symbol) | Leaves in rosettes on short (spur) shoots | (symbol) | Found in open countryside |
| (symbol) | Needles in fascicles of 2 | (symbol) | Found by water or on wet or moist sites |
| (symbol) | Needles in fascicles of 3 | ♂ male | ♀ female |

**(1)** – **(10)**   Zones of distribution (see p5)

## Abbreviations

| | | | | | | | |
|---|---|---|---|---|---|---|---|
| alt | alternate | lf | leaf | | | sim | similar |
| br | branch | lflt | leaflet | | | sp(p) | species |
| c. | about | lvs | leaves | | | | (plural) |
| cm | centimeter | m | meter | | | uns | underside |
| cv | cultivar | mm | millimeter | | | ups | upperside |
| fl | flower | opp | opposite | | | var | variety |
| fr | fruit | sev | several | | | vn | vein |
| infl | inflorescence | sh | shoot | | | yr | year |

4

# Tree distribution

Different trees have adapted to live in the wide range of
conditions to be encountered in North America, from Arctic to
subtropical, and climate is the main factor governing their
distribution. Altitude and land mass are also important: the same
trees may be found at high altitudes in the Appalachians as grow
near sea level in Canada, while the Great Lakes create a climate
warmer than if the region were all land. Atmospheric conditions
affect habitat locally—some trees can tolerate near-desert heat
and aridity while others require heavy rainfall or high humidity.
So a broad north–south scale of temperature is inadequate. In the
case of the west coast, for instance, the warm ocean and the
Rocky Mountains affect temperature and rainfall so that greater
hardiness is required for a tree to survive as one moves farther
inland, i.e. eastward, and not southward.

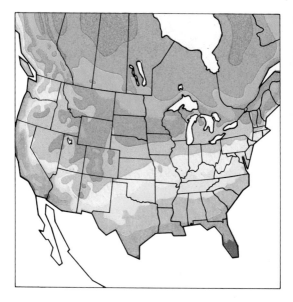

Average minimum annual temperature (°F)

| | | | |
|---|---|---|---|
| **Zone 1** below −50° | | **Zone 6** −10° to 0° | |
| **Zone 2** −50° to −40° | | **Zone 7** 0° to 10° | |
| **Zone 3** −40° to −30° | | **Zone 8** 10° to 20° | |
| **Zone 4** −30° to −20° | | **Zone 9** 20° to 30° | |
| **Zone 5** −20° to −10° | | **Zone 10** 30° to 40° | |

## Zones of tree hardiness

The widely recognized climatic zones on the map above,
although not absolute, give a good indication of plant hardiness.
Generally, a tree can also survive in zones warmer (higher
numbered) than its own. It is the minimum temperature which is
important—Great Britain, thought of as a colder country than
most of the USA, is covered by zones 8 and 9 thanks to the
temperate winters of its oceanic climate.

# How to identify trees

The identification of trees is a matter of putting together all the various pieces of information provided by the plant, and not just some of them. It is very tempting to latch on to some immediately striking feature and go no further, in which case you might well end up wrongly identifying a tree, as your only piece of evidence was misleading. For instance, you might know that ashes are generally characterized by pinnate leaves, and so never entertain the possibility that the tree in front of you with single leaves is an ash—but there is an exception to the rule, the Single-leaved ash. So whenever possible examine the foliage, buds, flowers, fruit, habit and bark, or as many as are visible at that time of the year.

The first step, before using the keys in this book, is to decide whether a tree belongs to the conifers or the broadleaf trees, something almost anybody can do. The actual difference is that the conifers have exposed ovules (which develop into seeds) whereas the broadleaf trees bear theirs enclosed in an ovary, but for identification purposes the differences indicated by the colloquial names of the groups are sufficient: the conifers bear cones, and have needle-like leaves which contrast with the wider foliage of the broadleaf trees.

Besides observing as many features as possible, bear in mind a few general points. Even within the plant, leaf character varies, the leaves at the top generally being smaller (although poplars are a notable exception). The number of lobes usually decreases with age, and a fine example of this is Holly. Remember that shoots of a tree that has been pruned will grow with abnormal vigor, giving rise to untypical proportions. As a rule you will find more typical features on short shoots. Features which generally occur in pairs can be found in threes or even fours, for example planetree maple fruits and the buds of Raywood ash.

## Foliage characteristics

Foliage should be examined thoroughly. First, important observations can be made at some distance from the tree itself, such as the color of the foliage, and whether or not it is pendent (hanging). With more detailed examination, leaves provide clues in variation of outline, and the way they are set upon the shoot. Other features to look at are the margins, the shape of the base and tip, the veins and the petiole, and the texture and hairiness.

**Conifers:** the illustrations below show four types of leaf arrangement. **Fascicles (a)**, borne by pines, are bunches of needles growing from one bud (needle is not a scientific term but simply a descriptive alternative to leaf). Leaves set in a **pectinate row (b)**, such as those of the yews and firs, are arranged in ranks along opposite sides of the shoot. **Awl-shaped leaves (c)**, as borne by the junipers, are set radially around the shoot. **Scale-like leaves (d)**, set densely round the shoot, are characteristic of Lawson cypress. See also the introduction to conifers, p 21.

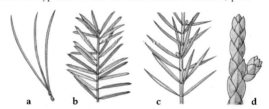

a      b      c      d

**Broadleaf trees:** the illustrations below show some basic leaf shapes of broadleaf trees, which have more varied foliage than the conifers. An important point to remember is that these are mostly deciduous, and are found in their mature form for a relatively short time.

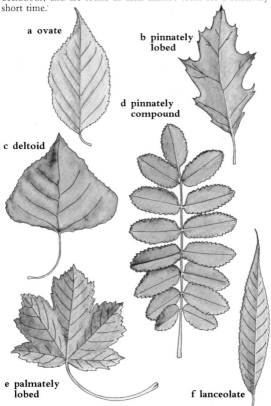

**a ovate**

**b pinnately lobed**

**d pinnately compound**

**c deltoid**

**e palmately lobed**

**f lanceolate**

Besides the basic shape, look at the **margins** (the edges of the leaf), which are serrated in **d** and **e** and entire in **f**; the **apex** (the end of the leaf farthest from the shoot), which is acute (or pointed) in **d** and acuminate (or needle-like) in **a**; and the **base** of the leaf, which is rounded in **a**, cuneate (or wedge-shaped) in **b** and **f**, truncate (or squared) in **c** and cordate (or heart-shaped) in **d**. See also the introduction to broadleaf trees, p 80.

**a**        **b**        **c**

The **petiole** or leaf stalk (above) can provide useful clues when its shape, particularly in cross-section, is examined. It is usually round **a** but can also be grooved **b**, as in Sweet cherry, or flattened **c**, as in Aspen. Some palms have stout hooked spines on the petioles. There is a bud at the base of the petiole, and usually the petiole is curved around it in a crescent shape. In some species, however, the bud is completely enclosed in the enlarged petiole base; when the leaf falls, a circular scar remains.

## Shoots and buds

**Shoots** are the new growth that springs from the buds at the start of the new season. They usually harden into their mature wooden state by midsummer and are then ready to bear the buds for the next season's growth.

Points to look for on shoots are the color, whether they are hairy or not, and their shape. Most shoots, when seen in cross-section, are round, but they can be angled or winged in some species. Cutting across a shoot reveals its basic structure, which is shown in illustration **c** below. In the center there is always the **pith**, sometimes a useful identification guide; it is generally a soft amorphous mass of various colors and texture, but is occasionally chambered, as in **b**.

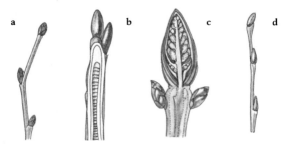

a        b        c        d

The **buds**, borne through the winter before producing the new growth, are a key identification feature. There is usually a single bud on the end of the shoot, known as the terminal bud, from which the shoot will continue to grow. Lateral buds are arranged along the sides of the shoot, and produce new lateral shoots or leaves. The way in which these lateral shoots are arranged can vary considerably, and the main differences ae illustrated above. They can be set alternately **a**, in opposite or nearly opposite pairs **b** and **c**, or in a spiral pattern **d**.

Buds are a useful identification guide where they are set in different ways; a few trees, such as the cypresses, do not have a terminal bud in winter, while oaks are distinctive through having a cluster of buds at the shoot tips. Many trees have buds laid down and covered by the bark; they remain dormant until the main shoot is damaged or more light reaches the bole, when they start growing and produce what are called epicormic shoots.

Other features to look for in buds are such obvious points as the shape, size and color, all of which differ widely. Buds may be ovoid, round, conic, cylindrical or spindle-shaped. Some species may have resinous or sticky buds **d**. One other key feature is the number of scales on the outside of each bud: some may be naked and have no scales at all **b**, or they may have one **c**, two **a** or many scales **d**, usually adpressed on top of each other.

a        b        c

## Thorns

Thorns are characteristic of a few species and occasional in others. The first and most obvious difference is the number in which they occur: the illustrations above show typical formations of thorns which grow in ones **a**, twos **b** or threes **c**. They are usually derived from shoots, and often carry buds at the base **a**.

## Bark

The bark, the protective outer covering of the stem, can be a useful identification feature, especially as it is available throughout the year. Although the young branches have smooth bark, that on the bole is usually split and broken as the diameter of the stem increases with age. It is the form that this breaking up takes which can be diagnostic, and the chief characteristics, such as color, are usually more pronounced in the bark higher up in the tree than in the mature bark near the base, because it has broken up more recently. Three basic types of bark are shown in the illustrations below: scaly fissures **a**, ridged and furrowed bark **b** and peeling bark **c**. The fissure color can be important.

**a** **b** **c**

# Glossary

**Adpressed** Closely pressed to an adjoining part of the plant, such as a bud against a twig or shoot (Hornbeam p101)

**Alternate** Leaves arranged singly along and on both sides of a stem

**Aril** Fleshy outer covering of a seed (Yew p24)

**Auricles** Ear-like appendages at the base of a leaf (Pedunculate Oak p106)

**Awl-like leaves** Leaves tapering to a slender, sharp point

**Axil** Upper angle formed between a twig and a shoot, or the junction between two veins

**Axil tufts** Pubescence in axil

**Bipinnate** Advanced type of pinnate leaf whose leaflets are themselves pinnate (Honey locust p156)

**Blade** Greatest part of a leaf, the part other than the leaf stalk

**Bloom** Powdery or waxy sheen of a shoot or leaf (Eucalyptus p189) which can easily be rubbed away

**Bole** Trunk or stem of a tree

**Bract** Modified leaf associated with a flower. In some trees they are more conspicuous than the flowers themselves (Dove tree p188); in conifers bract scales often extend beyond the edge of the fertile scales (p21)

**Buttress** Strengthened part of a bole or root that assists in supporting a tree (Beech p102)

**Calyx** Green cup of sepals, outside the petals, supporting the flower

**Carpel** Single unit of a flower comprising the stigma, style and ovary

**Catkin** Dense, usually long and pendent, group of flower bracts

**Chambered** Pith of a shoot that is broken up by hollow spaces

**Ciliate** Leaves having a hairy fringe to their margins

**Columnar-conic** Shape of a habit which is narrow and straight sided at its base and then becomes pyramidal

**Compound** Leaves comprising several separate leaflets

**Coppicing** The practice of cutting back trees such as willows to their stumps in order to promote growth

**Cordate** Heart-shaped; used especially of leaves (Linden p183) where the leaf bases curve away from the leaf stalk

**Corymb** Domed or flat-topped flower cluster with the external flowers opening first (p81)

**Crenate** Serration which has broad, rounded teeth

9

**Crown** Upper part of the tree; the branches and foliage

**Cultivar** Variety of tree selected in cultivation

**Cuneate** Wedge-shaped base of a leaf tapering into leaf stalk (Cork oak p118)

**Cupule** Circle of bracts which become cup-shaped and enclose fruit such as an acorn

**Cyme** Flat-topped flower cluster, the central flowers opening first (p81)

**Dioecious** Having male and female flowers on separate trees

**Drupe** Fruit containing a seed in the form of a stone

**Entire** Margin without either teeth or lobules

**Exserted** Extending

**Fascicle** Bundle or cluster (p5)

**Fastigiate** Tree habit which has very upswept, almost erect branches (Lombardy poplar p86, Cypress oak p106)

**Fluted** With alternate ridges and grooves (Dawn redwood p40)

**Glabrous** Smooth, not hairy

**Glaucous** Covered with a blue-grey or whitish bloom

**Globose** Irregularly spherical

**Habit** General appearance of a tree, usually from a distance

**Hilum** Paler, basal seed scar of a nut (Yellow buckeye p181)

**Hybrid** Species arising from the cross fertilization of two members of the same genus in which case the generic and specific names are separated by an 'x' (*Tilia x euchlora*) or (far less commonly) the members of two genera when the 'x' usually precedes the generic name (x *Cupressocyparis leylandii*)

**Illobulate** Entire, not bearing lobes

**Imbricate** Overlapping like roof tiles

**Impressed vein** Vein set below surface of leaf (American hornbeam p101)

**Inflorescence** Floral part of a plant

**Involucre** Circle of bracts developed to cover a fruit (Beech p102)

**Lanceolate** Shaped like the blade of a spear, as most willow leaves (Black willow p90)

**Layering** The ability of a tree to regenerate by a branch touching the ground where it roots and forms a new plant

**Leader** Leading shoot of tree and so one of the youngest; may droop and thereby be an identification feature

**Leaf scar** Imprint left on a twig or shoot of a fallen leaf. Also known as "leaf trace"

**Lenticel** Raised corky growth on twig or shoot which admits air to the interior of the branch

**Lobe** Rounded section of a leaf, divided by sinuses

**Lobulate** Bearing small lobes

**Margin** Leaf edge

**Midrib** Central vein of a leaf

**Monoecious** Having both male and female flowers on the same tree

**Oblique** Unequal sided base of a leaf, as in most elms (p121)

**Obovate** Ovate with the widest part beyond the middle of the leaf

**Obovoid** Ovoid and largest beyond the middle

**Operculum** Cap of fused petals that covers flower bud of Eucalyptus (p189), falls as flower opens

**Opposite** Set in pairs at the same level each side of the twig or shoot

**Orbicular** Almost circular shaped leaf

**Ovate** Leaf that is egg-shaped in outline, widest below the middle (p7)

**Ovule** Part of the plant that becomes the seed after fertilization

**Palmate** Leaf that has lobes or leaflets radiating from one simple point, like the fingers of a hand (p7); also applied to a leaf's venation pattern

**Panicle** Compound inflorescence whose flowers branch from a central stem

10

**Pectinate** 2-ranked arrangement of leaves either side of a central shoot in the nature of the teeth of a comb (p5)

**Peduncle** Flower stalk

**Peltate** Shield-shaped

**Perfect** Having male and female organs combined in one flower and not as separate flowers

**Petiole** Leaf stalk

**Petiolule** The stalk attaching leaflets to a central rachis

**Pinna** Leaflet or primary division of a pinnate leaf

**Pinnate** Compound leaf with leaflets arranged regularly each side of a rachis (p7)

**Pith** Softer, central section of a shoot's stem

**Pneumataphores** Special aerial root of Bald cypress (p39) that provides air to roots on waterlogged sites

**Pollarding** The practice of lopping trees at about 3 m above the ground in order to encourage further growth

**Pome** Fruit of several carpels enclosed in thick flesh (Crab apple p140)

**Pubescence** Covering of soft and short hairs

**Pulvinus** Swelling on some conifer shoots from which each leaf grows (p42)

**Raceme** Inflorescence of stalked flowers growing from a rachis (p6)

**Rachis** Central stalk of a compound (pinnate) leaf or inflorescence

**Samara** Winged fruit (Maple p172)

**Serrate** Toothed

**Sessile** Non-stalked

**Sheath** Tubular envelope enclosing a bunch of fascicle leaves (Pine p42)

**Simple** Leaf consisting of one single blade (Tulip tree p134)

**Sinuate** Strongly waved

**Sinus** Recess between lobes

**Stipule** Appendage usually at the base of a petiole

**Stomata** Orifices in a leaf used to let air into the leaf

**Subsessile** With a minute stalk

**Ternate** Set in threes

**Tomentose** Having a dense, woolly pubescence

**Trifoliate** Leaf comprising 3 leaflets

**Truncate** Abrupt end to a leaf base or tip

**Umbel** Inflorescence with pedicels all arising from the same point (p81)

**Umbo** Raised centre of the scale of a pine cone

**Valve** Section into which a fruit capsule splits

**Venation** Pattern of veins

**Vein axils** Upper angle between two veins

# Identification keys

Don't shy away from botanical keys; they are a tried and tested aid to tree identification. The keys to conifers, broadleaf trees and leafless winter shoots included here will lead you to the genus of the tree concerned; from the page numbers given you can quickly locate the species in question.

To use these keys, select a good specimen and progress through the listed options the key offers. If the answer to the first part of a choice is "yes", move on to the name or number indicated; if "no" when only one choice is offered, move to the next number. Eventually you will reach the genus (or selection of genera) to which your species belongs. In winter, there is nothing to prevent you using the broadleaf key on fallen or dried leaves or using the leafless shoot key while the shoots are still leafy.

Learn to use the keys with a species already familiar to you. With practise you will find that their use together with detailed observation in the field will greatly improve your knowledge of trees and your ability to identify them.

## Preliminary key

**1**  **a** Tree in leaf    ▷ **2**
    **b** Tree not in leaf    **Key C**

**2**  **a** Leaf needle- or scale-like; veins absent or (more rarely) parallel    **Key A**
    **b** Main veins branched, reticular    **Key B**
    **c** Leaf veins parallel, leaves more than 20 cm    **Key D**

### Key A Conifer trees

**1**  **a** Lvs broad (over 4 mm), vns parallel    ▷ **2**
    **b** Lvs narrow or scale-like    ▷ **3**

**2**  **a** Lvs broadest at base, hard, sharp ***Araucaria*** 26
    **b** Lvs broadest at apex, soft, lobed, rounded ***Gingko*** 23

**3**  **a** Lvs set singly, in opp prs or whorls of 3    ▷ **4**
    **b** Lvs set spirally, singly or in bundles of 2–5    ▷ **8**

**4**  **a** Lvs linear, in fern-like fronds; new buds set below side shoots ***Metasequoia*** 40
    **b** Lvs scale- or awl-like    ▷ **5**

**5**  **a** Fruit berry-like; lvs set in 2s or 3s ***Juniperus*** 36
    **b** Fr a cone, lvs always paired    ▷ **6**

**6**  **a** Foliage in flat sprays    ▷ **7**
    **b** Foliage in 3-D sprays ***Cupressus*** 27

**7**  **a** Cone rounded x ***Cupressocyparis*** 29 or ***Chamaecyparis*** 30
    **b** Cone flask-shaped ***Thuja/Biota*** 34, ***Thujopsis*** 35 or ***Calocedrus*** 36

**8**  Lvs long (10–15 cm), in spaced whorls ***Sciadopitys*** 41

**9**  **a** Lvs thin-textured, set in fern-like fronds ***Taxodium*** 39
    **b** Lvs not as above    ▷ **10**

**10**  **a** Sh not green (except when very young)    ▷ **11**
    **b** Sh green or yellow-green for at least 2 yrs    ▷ **16**

**11**  Lvs in bundles of 2, 3 or 5 ***Pinus*** 64

**12**  **a** Lvs in rosettes on 2nd yr and older shs    ▷ **13**
    **b** Lvs always single, spaced along shs    ▷ **14**

**13** **a** Lvs soft, turning yellow in autumn; cones have persistent, open scales *Larix* 54
**b** Lvs hard; cones with deciduous scales *Cedrus* 52

**14** Lvs set on prominent pegs projecting from shoot *Picea* 58 or (more rarely) *Tsuga* 62

**15** **a** Cones small (under 3 cm), pendulous; buds very small (under 2 mm), rounded *Tsuga* 62
**b** Cones over 3 cm, pendulous, persistent, with exserted, trident bracts; *Pseudotsuga* 61
**c** Cones erect, over 5 cm, with deciduous scales *Abies* 43

**16** **a** Lvs flat in x-section, parted on sh ▷ 17
**b** Lvs round in x-section ▷ 19

**17** Bark thick, very soft *Sequoia* 39

**18** **a** Lvs spine-tipped, very sharp to touch *Torreya* 24
**b** Lvs softer *Taxus* 24, *Podocarpus* 25 or *Saxegothaea* 25

**19** **a** Lvs short (4–7 mm), pointed, dotted with stomata; bark thick, soft *Sequoiadendron* 38
**b** Lvs longer (to 1.5 cm), less adpressed; bark thin, hard, stringy *Cryptomeria* 41

### Key B Broadleaf trees
An asterisk alongside the generic name indicates that at least some species in the genus are evergreen

**1** **a** Lvs in opp or nearly opp prs on sh ▷ 2
**b** Lvs alt on sh ▷ 12
**c** Lvs in whorls of 3, or both opp and alt *Lagerstroemia* 196, *Fraxinus* 197 (rarely), *Catalpa* 205 or *Chilopsis* 206

**2** **a** Lvs compound ▷ 3
**b** Lvs simple ▷ 4

**3** **a** Lvs pinnately compound *Euodia* 168, *Acer* 171 or *Fraxinus* 197
**b** Lvs palmately compound *Acer* (rarely) 177 or *Aesculus* 180

**4** **a** Lvs toothed or lobed ▷ 5
**b** Lf margins entire (or with 1–2 prs of large teeth) ▷ 8

**5** Lvs lobed or lobulate *Acer★* 171

13

**6** Lvs thick, leathery, dark green
*Phillyrea*★ 204

**7 a** Lvs narrowly oval; sh squarish
*Euonymus* 168
**b** Lvs ovate with cordate bases; sh
rounded, brown *Cercidiphyllum* 130
**c** Lvs elliptic, finely toothed
*Forestiera* 203

**8 a** Lvs leathery, evergreen, glaucous
or glossy above ▷ **9**
**b** Lvs not leathery, deciduous ▷ **11**

**9 a** Lvs willow-like, felted white
below *Olea*★ 202
**b** Lvs glabrous ▷ **10**

**10 a** Lvs same both sides, glaucous
*Eucalyptus*★ 189
**b** Lvs not same both surfaces,
glossy above *Ligustrum*★ 202

**11 a** Lvs less than 10 cm, veins curve
parallel to margins *Cornus* 191
**b** Lvs more than 10 cm *Paulownia*
204, *Catalpa* 205 or *Chionanthus* 203

**12 a** Lvs simple ▷ **21**
**b** Lvs pinnate ▷ **13**
**c** Lvs bipinnate *Gleditsia* 156,
*Acacia*★ 157, *Koelreuteria* 186,
*Melia* 162, *Gymnocladus* 159,
*Prosopis* 158 or *Delonix* 158

**13** Lvs always comprise 3 lflts
*Laburnum* 160

**14** Pith chambered, not solid *Juglans* 91

**15** Lflts alt on rachis *Cladrastis* 159

**16** Terminal lflt or 3 terminal lflts
largest *Carya* 94

**17 a** Bud hidden in base of petiole ▷ **18**
**b** Bud not so hidden ▷ **19**

**18 a** Sh round, smooth, green, yellow
or paler *Gleditsia* 156, *Sophora* 157
or *Rhus* 163
**b** Sh ridged, brown *Robinia* 156

**19 a** Lflts entire or with 1–3 prs of
teeth at base *Ailanthus* 161, *Rhus*
163 or *Pistacia*★ 163
**b** Lflts toothed ▷ **20**

**20 a** Toothing regular *Sorbus* 139 or
*Rhus* 163
**b** Toothing very coarse
*Koelreuteria* 186

**21** Lvs have truncate bases, indented apexes *Liriodendron* 134

**22** Sh green for several yrs, often spined, petiole usually broad-winged, fr orange-like *Citrus*★ 165 or *Fortunella*★ 167

**23** Lvs concentrated at sh tip, fewer along sh; fruit an acorn *Quercus*★ 105

**24** Bk in upper crown smooth; peeling orange and yellow-pink *Arbutus*★ 193

**25** Shs with 1.3 cm spines, lvs with milky sap *Maclura* 129

**26** **a** Lf margins entire, toothless ▷ 24
   **b** Lvs toothed or lobed ▷ 31

**27** **a** Lvs same both sides, pendulous *Eucalyptus*★ 189

**28** Shs and willow-like lvs silvery-white with dense scales *Elaeagnus* 190

**29** Bk brown, flaking to yellow and pink *Parrotia* 136

**30** **a** Lf margins wavy or crinkled ▷ 27
   **b** Lf margins flat ▷ 28

**31** **a** Lvs very glossy, dark green above *Diospyros* 194
   **b** Lvs pale green, slightly glossy *Pittosporum*★ 179
   **c** Veins impressed, pubescent below *Salix* 88

**32** Lvs orbicular *Cercis* 155 or *Coccolobis*★ 130

**33** **a** Lvs have vns curving parallel to margin *Cornus* 191

**34** Lvs prominently 3-vnd at or nr base *Celtis* 128 or *Sassafras* 135

**35** **a** Lvs elliptic to obovate *Salix* 88, *Magnolia* 131, *Persea*★ 134, *Anacardium*★ 162 or *Nyssa* 187
   **b** Lvs lanceolate *Pyrus* 133
   **c** Lvs elliptic to lanceolate, evergreen, with strong spicy odor when crushed *Umbellularia*★ 135

**36** **a** Lvs lobed ▷ 32
   **b** Lvs toothed or lobulate, not lobed ▷ 36

15

**37** Lvs 50 cm on stout hollow petioles *Carica*★ 131

**38 a** Vns, arise at lf base (palmate vns) ▷ **33**
**b** Vns arise along midrib (pinnate vns) *Crataegus* 138 or *Sorbus* 139

**39** Bud hidden in hollow base of petiole *Platanus* 169

**40** Lf uns densely hairy *Populus* 83

**41 a** Sinuses obtuse, rounded *Morus* 128 or *Ficus* 129
**b** Sinuses acute *Liquidambar* 136

**42** Margins have spine-like teeth *Castanea* 120, *Ilex*★ 164 or *Prunus*★ 151

**43** Sh has bristly, glandular hrs *Corylus* 99

**44 a** Lvs rounded ovate, more than 4 cm, bases cordate or oblique *Tilia* 182 or *Davidia* 188
**b** Lvs less than 4 cm, or not as above ▷ **39**

**45** Lvs deltoid or orbicular; petiole flattened or lf uns "painted" white *Populus* 83

**46 a** Pith finely chambered, fr 4-winged *Halesia* 196
**b** Pith solid ▷ **47**

**47 a** Vns deeply impressed *Salix* 88, *Ostrya* 100, *Carpinus* 101, *Nothofagus* 119 or *Parrotia* 136
**b** Vns not deeply impressed ▷ **48**

**48** Sh green for 1–2 yrs or petiole has 2 (1–4) glands near lf blade *Prunus* 144

**49** Lvs lanceolate *Salix* 88

**50** Lf uns has silvery white pubescence *Sorbus* 139 or *Malus* 140

**51** Margins crinkled or wavy *Alnus* 98, *Fagus* 102, *Stewartia* 187 or *Styrax* 195

**52** Lvs obliquely based, doubly serrate *Ulmus* 121

**53** Fruit a persistent, ovoid, woody cone-like catkin *Alnus* 98

**54** Fruit a cylindrical or ovoid catkin, deciduous *Betula* 95

**55** **a** Lvs dark, shiny green, elliptic-oblong, fruit strawberry-like ***Arbutus*** ★ 193

**56** Lvs shiny above, leathery, oval to ovate ***Camellia*** ★ 186

**57** Lvs elliptic, fr a red berry ***Ilex*** ★ 164

**58** Teeth deep, single, in 12 or less prs ***Zelkova*** 127 or ***Ulmus*** 121

**59** Bark silvery-gray, fluted; lf vns parallel, impressed ***Carpinus*** 101

**60** Shoots have spines; fruit berry-like ***Crataegus*** 138

**61** **a** Flowers and fruits in umbels ***Malus*** 140
**b** Flowers and fruits in racemes ▷ 60

**62** Lvs nearly round or elliptic, densely pubescent and glaucous below **Styrax** 195

**63** **a** Lvs under 5 cm, dull; fls in spring ***Amelanchier*** 137
**b** Lvs over 5 cm, glossy; fls in autumn ***Oxydendrum*** 194

## Key C Leafless winter shoots

**1** **a** Buds in opp prs or whorls of 3 ▷ 2
**b** Buds alt along sh ▷ 11

**2** Lf scars set above buds; bark fibrous, red-brown ***Metasequoia*** 40

**3** **a** Buds with 2 outer scales ▷ 4
**b** Buds with sev outer scales or none ▷ 5

**4** **a** Buds elongated ***Cornus*** 191
**b** Buds stalked ***Acer*** 171
**c** Buds pointed; sh winged or ribbed ***Lagerstroemia*** 196

**5** Sh squarish, green for 1st winter ***Euonymus*** 168

**6** **a** Terminal bud lacking ▷ 7
**b** Terminal bud present ▷ 8

**7** **a** Sh slender ▷ 8
**b** Sh stouter, over 5 mm ***Paulownia*** 204 or ***Catalpa*** 205

**8** **a** Buds 3–6 mm, crimson-brown, glabrous ***Cercidiphyllum*** 130
**b** Buds with several imbricate scales, rusty hrd, both opp and alt on sh ***Chilopsis*** 206

**9** Buds naked *Euodia* 168

**10 a** Buds have 4–8 pubescent scales *Fraxinus* 197, *Forestiera* 203 or *Chionanthus* 203
   **b** Buds have 8+ scales ▷ **10**

**11 a** Sh stout, buds large *Aesculus* 180
   **b** Sh less stout, buds under 8 mm, not resinous *Acer* 171

**12 a** Shs have spines ▷ **13**
   **b** Shs do not have spines ▷ **14**

**13 a** Spines brown, paired either side of bud *Robinia* 156 or *Prosopis* 158
   **b** Spines stout, 3-pronged, often on bole *Gleditsia* 156
   **c** Spines single with buds set at bases *Maclura* 129, *Crataegus* 138, *Malus* 140, *Pyrus* 143 or *Prunus* 144 (rarely in most except *Maclura* and *Crataegus*)

**14 a** ♂ catkins exposed over winter at twig end ▷ **15**
   **b** Not as above ▷ **18**

**15** Bark shaggy, broken into many small plates *Ostrya* 100

**16** Buds stalked; fr a woody, cone-like catkin, persistent *Alnus* 98

**17 a** Sh has bristly hrs; fr a nut *Corylus* 100
   **b** Sh has no bristly hrs; fr a catkin *Betula* 95

**18** Fr a hard, bony, seeded drupe *Melia* 162

**19 a** Buds naked ▷ **20**
   **b** Buds have scales ▷ **22**

**20 a** Sh ribbed, brown *Robinia* 156
   **b** Sh round ▷ **21**

**21 a** Sh green *Gleditsia* 156 or *Cladrastis* 159
   **b** Sh brown or brownish *Rhus* 163

**22 a** Buds have single scale ▷ **23**
   **b** Buds with 2+ scales ▷ **26**

**23 a** Lf scar around bud ▷ **24**
   **b** Lf scar only below bud ▷ **25**

**24 a** Bud conic, single *Platanus* 169
   **b** Buds adpressed, usually 2+ together *Styrax* 195

18

**25 a** Terminal bud large, pinched at base *Magnolia* 131
**b** Terminal lacking, other buds conic, adpressed *Salix* 88

**26** Winter buds clustered at sh end; on rest of sh are spaced, fewer, smaller *Quercus* 105

**27 a** Bud single, central on sh short ▷ **28**
**b** Buds not as above ▷ **29**

**28 a** Sh ribbed; fr a persistent cone *Larix* 54
**b** Sh round; fr a drupe, shed in autumn *Gingko* 23

**29** Bud spindle-shaped, pointed, 2 cm *Fagus* 102

**30** Fr 4-winged *Halesia* 196

**31** Pith chambered *Juglans* 91

**32** Bud acute, adpressed; sh slender, zig-zagged *Carpinus* 101 or *Nothofagus* 119

**33** Bud resinous *Populus* 83 or *Sorbus* 139

**34 a** Bud stalked ▷ **35**
**b** Bud sessile ▷ **36**

**35 a** Bark flakes to pink or yellow-green below *Parrotia* 136
**b** Bark tight, regularly ridged *Liriodendron* 134

**36 a** Sh green or greenish ▷ **37**
**b** Sh not as above ▷ **42**

**37 a** Sh green for 2+ years ▷ **38**
**b** Sh partially green or green for only 1 yr ▷ **40**

**38** Lf scar around bud *Gleditsia* 156, *Sophora* 157 or *Cladrastris* 159

**39 a** Lf buds with 2 unequal scales *Tilia* 182
**b** Lf buds with 2+ scales *Laburnum* 160 or *Sassafras* 135

**40 a** Bud large, over 5 mm *Carya* 93 or *Ficus* 129
**b** Bud under 5 mm ▷ **41**

**41 a** Bud yellow-green *Oxydendrum* 194
**b** Bud red-brown *Nyssa* 187
**c** Not as above *Prunus* 144

**42** **a** Shs angled *Castanea* 120
**b** Shs with corky ridges or wings *Liquidambar* 136 or *Ulmus* 121
**c** Sh round ▷ **43**

**43** **a** Sh over 5 mm ▷ **44**
**b** Sh under 5 mm ▷ **45**

**44** **a** Sh bloomed for 1+ yrs *Gymnocladus* 159
**b** Sh not bloomed *Ailanthus* 161

**45** Sh silvery with dense scales in 1st winter *Elaeagnus* 190

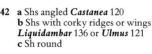

**46** **a** Buds with 2 or 3 scales ▷ **47**
**b** Buds with 4+ scales ▷ **49**

**47** Buds with 3 scales; bark black, fissured into squares *Diospyros* 194

**48** **a** Sh pale reddish or coppery *Koelreuteria* 186
**b** Sh zig-zagged, bud with 1 large and 1 small scale *Tilia* 182

**49** Bud ovoid, pointed, shiny red or chestnut brown *Morus* 128, *Sorbus* 139, *Pyrus* 143, *Prunus* 144, *Davidia* 188

**50** **a** Buds in 2s or more above lf scar ▷ **51**
**b** Buds set singly above lf scar ▷ **52**

**51** **a** Bud silky; bark flaking *Stewartia* 187
**b** Bark dark red; fr a pod hanging from old wood *Cercis* 155

**52** Bark red brown, stringy; twigs reddish, supple *Taxodium* 39

**53** Bud small, less than 2 mm *Ulmus* 120, *Zelkova* 127, *Halesia* 196 or *Celtis* 128

**54** **a** Bud scaled, with whitish hrs *Populus* 83, *Malus* 140 or *Sorbus* 139

### Key D Palm key

**1** **a** Leaves pinnate *Phoenix* 207 or *Roystonea* 207
**b** Leaves palmate ▷ **2**
**c** Leaves needle-like *Cordyline* 211 or *Yucca* 211

**2** **a** Petioles with stout teeth *Trachycarpus* 206, *Washingtonia* 209 or *Chamaerops* 208
**b** Petioles not toothed *Thrinax* 210 or *Sabal* 210

# The conifers

The Gymnospermae, known colloquially as the conifers, belong to three different botanical groups or orders, the Gingko group, the Yew group and the true conifers. There are some 50 genera of conifers which embrace over 600 species; many of these are native to North America, and other conifers which are now familiar have been introduced from Europe and Asia. The simple upright habit, making good timber, and the fact that they can grow fast on poor soils and in harsh climates have made conifers ideal for widespread forestry cultivation.

The fundamental botanical difference between conifers and all others plants is that their ovules, which develop into seeds, are carried naked. This gives the group their scientific name, the Gymnospermae, and distinguishes them from the Angiospermae or "enclosed ovule" trees, known as broadleaf trees. The ovules are borne on the scales of the female flower, an immature cone, which closes its scales after fertilization.

This is of little help with identification, however, and it is far better to examine other characteristics of growth, habit and fruit. Conifers in general have a strongly monopodal growth habit, with a single stem and much lighter, smaller side branches. Some do not maintain this character into old age and particularly in European larch and some of the firs one or two very heavy horizontal side branches may turn up at the ends and form competing leaders. Some of the yews also have a tendency to heavier branching; forking often occurs and one of the stems outpaces and suppresses the other.

a       b       c

The illustrations above show some points to look for when examining the superficially similar needles of most conifers. The spruce needle **a** has a pointed apex, is squarish in section, and is similar in appearance above and below. The Silver fir needle **b** and **c** is notched at the apex, much flatter, and has white bands on the underside which are in fact the stomata, pores which open and close as the light intensity and humidity vary. Conifer leaves are generally· arranged helically along the shoot, although this is sometimes disguised when the leaves are twisted at the base and appear to be arranged pectinately, spreading either side of the shoot. In some species the leaves may be arranged on short shoots, most notably amongst the cedars, larches and gingko. Although still in a helix the leaves are compressed and appear to be a whorl with a central bud.

**The seeds are ripe** in the autumn six or eighteen months are fertilization, and the cone scales usually open, in dry weather, to release them on the wind. The seeds are tucked between the fertile and bracts scales (left).

They are usually winged to assist in their distribution. The number of seeds per scale can vary from one to as many as twenty. In some conifers the seeds remain attached to the scales, which themselves separate from the cone and fall.

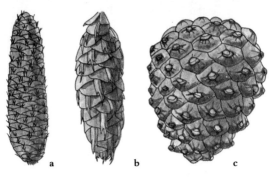

a        b        c

The colloquial name of the conifers derives from the cones, the hard woody fruits with overlapping scales which bear their seeds. They vary enormously from one genus to another, and are perhaps the best means of identification; the illustrations on this page show some points to look for. The fir cone **a** always sits erect on the shoot, and is long and cylindrical. It is also distinctive because of the bracts, which are reflexed, and stand out from the body of the cone. This is one of the cones which is rarely found whole on the ground, because their scales are deciduous, falling to the ground and leaving a long bare stalk or "candle" on the tree. The Douglas fir cone **b** is equally distinctive in its growth, as it always hangs from the shoot. It is also shorter and more ovoid in shape, and has prominent bracts, which are three pointed. The pine cone **c**, with its rounded base, pointed tip and triangular scales, is often imagined to be typical of all cones, although it only reaches this form after eighteen months. A feature of these cones is the blunt projection known as an umbo borne on the outside of each scale. Pine cones are often asymmetrical, something comparatively rare in other genera.

d        e        f

The Juniper fruit **d** is quite unlike that of any other conifer, in that it is not really a cone at all but a soft fleshy berry. As the fruit takes two or three years to ripen the green immature fruit and the ripe blue ones can usually be found on the same tree. The female flowers begin life like all other conifers, as open scales, but they mature into hard round berries, recognizable by the scars and blunt points left from what were their scales. The Western red cedar cone **e** is unusual in not having overlapping scales, but ones which separate from the base. The center of the cone is leathery and ovoid, visible behind the few spreading scales. The cone of the Coast redwood **f** is much smaller and very knobbly, with the scales presenting a diamond-shaped surface. These scales each have a central hollow, and as they ripen from green to brown they shrink and separate. The seeds, which can easily be obtained by shaking a newly ripened cone, vary considerably in size and appearance. The conifers usually bear their seeds two per scale.

# Gingko family Gingkoaceae

## Gingko

### *Gingko biloba*

**30 m. Crown** *columnar, broadening with age.* **Branches** *short, numerous, dipping when old.* **Bark** *ridged, fissured*

**Buds** *set spirally around shoots*

**Leaves** *variable, from 6–12 cm, set in whorls of 2–5 on short, slow growing, older shoots; larger, set singly on new shoots.* **Blades** *open pale yellow-green, golden in autumn*

**Leaf** *ribbed, divided into 2 or more lobes on soft shoots; those on long shoots may be undivided*

**Veins** *straight, parallel, non-dividing*

The Gingko is the only survivor of a group of trees that flourished some 200 million years ago. Its primitive ancestry shows in its regular, dichotomous (forked) venation and the rudimentary method by which its ovule is fertilized. As in the ferns, this is by free-swimming sperm cells and fertilization often occurs *after* its ovoid, yellow fruit, found only on female trees, has fallen. This emits a putrid stench once its fleshy coat begins to rot. The Gingko is only native to a remote part of China where it was adopted as a sacred tree by Buddhist monks who carried it to Japan. From there it was introduced to Europe and America in the eighteenth century. Specimens planted then still survive and an immunity to air pollution and most pests and diseases makes Gingko ideal for towns. Its name, "Maidenhair tree", is based on the similarity of its leaves to those of Maidenhair fern.

# Yew family Taxaceae

Yews are usually dioecious; females produce solitary seeds in a fleshy aril. Leaves are set spirally or in pectinate ranks.

## English yew

### *Taxus baccata*

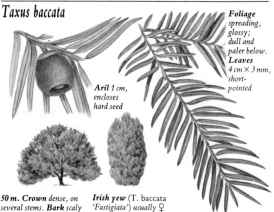

**Foliage** *spreading, glossy; dull and paler below.* **Leaves** *4 cm × 3 mm, short-pointed*

**Aril** *1 cm, encloses hard seed*

**50 m. Crown** *dense, on several stems.* **Bark** *scaly*

**Irish yew** (T. baccata 'Fastigiata') *usually* ♀

A long-lived ornamental evergreen, English yew is widely grown but replaced from New England northward, where it is not hardy, by its hybrid *T. x media*, with sparser, longer foliage, and 'Hicksii', a small tree similar to Irish yew in habit.

## California nutmeg

### *Torreya californica*

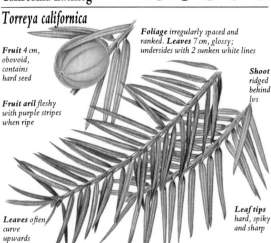

**Foliage** *irregularly spaced and ranked.* **Leaves** *7 cm, glossy; undersides with 2 sunken white lines*

**Fruit** *4 cm, obovoid, contains hard seed*

**Shoot** *ridged behind lvs*

**Fruit aril** *fleshy with purple stripes when ripe*

**Leaves** *often curve upwards*

**Leaf tips** *hard, spiky and sharp*

An open-crowned tree with spreading branches and stout shoots, this species reaches 15 meters and its fruit has seeds resembling those of the true nutmeg. Japanese nutmeg (*T. nucifera*) has shorter and decurrent needles.

# Podocarp family Podocarpaceae

These species come mainly from the southern hemisphere and carry cones with fleshy scales and less regular foliage than the yew family. Male and female flowers may appear on the same tree.

## Prince Albert yew

### Saxegothaea conspicua

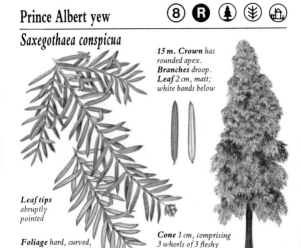

**15 m. Crown** has rounded apex. **Branches** droop. **Leaf** 2 cm, matt; white bands below

**Leaf tips** abruptly pointed

**Foliage** hard, curved, set closely but unevenly

**Cone** 1 cm, comprising 3 whorls of 3 fleshy scales. **Bole** often bends

A Chilean tree which was named after Queen Victoria's consort, this differs from English yew in its fruit, irregular foliage and leaf undersides while its less dense, pendulous shoots distinguish it from *P. andinus*.

## Chile yew

### Podocarpus andinus

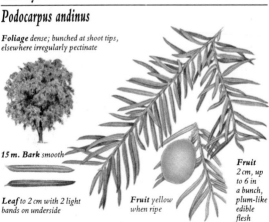

**Foliage** dense; bunched at shoot tips, elsewhere irregularly pectinate

**15 m. Bark** smooth

**Leaf** to 2 cm with 2 light bands on underside

**Fruit** yellow when ripe

**Fruit** 2 cm, up to 6 in a bunch, plum-like edible flesh

Chile yew is a small tree of variable habit which is sometimes confused with Prince Albert yew (above). Large-leaf podocarp (*P. macrophylla*), native to China and Japan, has 5–10 cm leaves, up to 1 cm broad, and a gray, shallowly fissured shredding bark.

# Chile pine family Araucariaceae

This family comprises two genera and 36 species, all natives of the southern hemisphere. Trees are either male or female, the female ones carrying globose or ovoid cones containing many scales, each with one seed. Leaves are hard with parallel venation and are usually broad and arranged spirally.

## Monkey puzzle

### *Araucaria araucana*

**25 m. Crown** domed or conical, variable in depth. **Bole** very straight with occasional suckers. **Bark** dark gray, fissured, with old branch scars

**Branches** in whorls, horizontal near apex of crown, drooping elsewhere, upturned at tips. **Foliage** remains green for 10–15 years, persisting after drying and turning brown

♂ **cone** 6–10 cm, green at first then brown, dry, papery, hanging in clusters of 1–6 near branch tips, persisting 1 year

**Leaves** 4 × 1 cm, set spirally, interlocking round shoot, hard, leathery, sharply tipped, with parallel veins

♀ **cone** large, 15 cm, ripens to brown in 2nd autumn when 4 cm seeds released as it breaks up

Indigenous to Chile where its tasty seeds were once an important food of the Araucano tribe, the Monkey puzzle was first introduced to America in 1795 and its common name, alluding to the problems its sharp foliage would give potential climbers, was first used in 1834. The Norfolk Island pine (*A. heterophylla*) has softer, awl-shaped leaves. It can grow in the open in warm areas but is more common as an indoor plant.

# Cypress family Cupressaceae

Trees in the Cypress family are distinguished from other conifers, except *Metasequoia* (p 40), by their paired or ternate leaves. Those of juvenile plants are always awl-shaped, a condition retained by some special and many cultivars. Adult leaves are mostly small, scale-like and adpressed. Members of Cupressaceae carry male and female flowers on the same or separate trees and three types of cone are produced: those of *Cupressus* (pp 27–8) and *Chamaecyparis* (pp 30–3) are globose or ellipsoid with peltate scales; those of *Juniperus* (pp 36–7) have unique fleshy scales which are fused together. All other Cypress genera—including *Thuja* (pp 34–5), *Thujopsis* (p 35) and *Calocedrus* (p 36)—have larger cones with woody scales hinged at their bases.

## Monterey cypress

⑦ 🌲 ⅄ 🚫 🏠

### *Cupressus macrocarpa*

**Young crown** columnar conic with erect leader

**'Lutea'** 26 m, perhaps hardier than the type. **Foliage** thicker, grows more slowly, greener when planted in shade

**35 m. Habit** usually this shape when mature but can broaden and become flat-topped

**Leaves** small, 2 mm, scale-like

**Branches** ascend at base, level near apex. **Bark** ridged

**Foliage** slender but set densely, lemon-scented when crushed. **Shoots** forward-pointing, almost cylindrical, covered by lvs

**Foliage** thicker, shorter on older trees; salt-resistant

**Cone** to 4 cm when mature, has 8 or 10 peltate scales

This species is found wild on the coast of California and was widely planted as a hedge tree although it has now been largely superseded in that role by the hardier and faster growing Leyland cypress (p 29). A fungal disease, *Seiridium* (= *Coryneum*) *cardinale*, attacks Monterey cypress as well as Italian cypress (p 28) and can prove fatal.

# Italian cypress

## *Cupressus sempervirens*

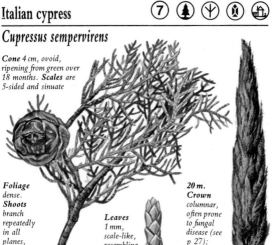

**Cone** 4 cm, ovoid, ripening from green over 18 months. **Scales** are 5-sided and sinuate

**Foliage** dense. **Shoots** branch repeatedly in all planes, unlike Chamaecyparis

**Leaves** 1 mm, scale-like, resembling C. macrocarpa but sharper, flatter

**20 m. Crown** columnar, often prone to fungal disease (see p 27); spreads wider with level brs in wild

This is the classic cypress, native to Greece and the Balkans and now widely distributed throughout Mediterranean countries. Cedar of Goa (*C. lusitanica*), from Mexico, has a broader crown, smaller, glaucous cones and spreading, pointed leaves.

# Smooth cypress

## *Cupressus glabra*

**Foliage** open, branching almost at 90°

**Bark** smooth, flaked, can be cherry-red

**Cone** 15–25 mm, often clustered and persistent, bloomed at first. **Scales** have low, curved spikes

**20 m. Crown** conical or ovoid, fairly dense with blunt apex. **Branches** ascend. **Branchlets** horizontal. **Bole** short

Distinguished by its bark, this tree is being increasingly planted as an ornamental. It was once classified as a cultivar of Arizona cypress (*C. arizonica*) whose bark is much rougher.

# Leyland cypress

## *Cupressocyparis leylandii*

**Cone** *(when present) 3 cm, globular, brown*

**Foliage** *dark green or gray, in irregular, flattish planes (with neither the truly flat sprays of* Chamaecyparis *nor the more radial arrangement of* Cupressus*); less dense near leader.* **Shoots** *branch repeatedly.* **Leaves** *slightly incurved with ridged glands.* **Growth** *fast, can exceed 1 m a year*

**40 m. Crown** *dense, columnar with conic or rounded conic apex.* **Branches** *ascend steeply.* **Leader** *leans slightly, not drooping as in Lawson cypress (pp 32–3).* **Bark** *initially smooth, then ridged becoming stringy*

*'Castlewellan' is extremely vigorous and its foliage, arranged in plumes, turns bronze-green in winter*

Leyland cypress is a naturally spontaneous hybrid of Monterey cypress (p 27) and Alaska cedar (p 31) and the qualities inherited from its parents—the vigorous growth rate of the first and the adaptable durability of the second—have made this tree unbeatable as a hedge conifer. Several clones are now marketed. The most popular, 'Haggerston Grey', was first raised in 1888 but two recently introduced golden forms, 'Castlewellan' and 'Robinson's Gold' are becoming increasingly acceptable. Leyland rarely sets viable seed but plants are easily raised from cuttings.

# Hinoki cypress

## *Chamaecyparis obtusa*

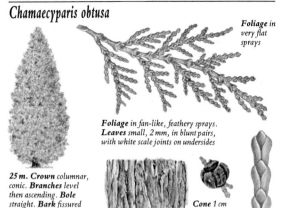

**Foliage** *in very flat sprays*

**Foliage** *in fan-like, feathery sprays.*
**Leaves** *small, 2 mm, in blunt pairs, with white scale joints on undersides*

**25 m. Crown** *columnar, conic.* **Branches** *level then ascending.* **Bole** *straight.* **Bark** *fissured*

**Cone** *1 cm*

This native of Japan can be distinguished from other "false" cypresses by its blunt, incurved leaves and larger, round cones. 'Crippsii', a popular cultivar, has bright, golden-yellow foliage which darkens to green inside the crown.

# Sawara cypress

## *Chamaecyparis pisifera*

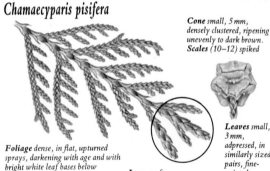

**Cone** *small, 5 mm, densely clustered, ripening unevenly to dark brown.* **Scales** *(10–12) spiked*

**Leaves** *small, 3 mm, adpressed, in similarly sized pairs, fine-pointed, incurved*

**Foliage** *dense, in flat, upturned sprays, darkening with age and with bright white leaf bases below*

**Leaves** *of many Sawara cultivars are always awl-shaped and set in pairs. Those of* **'Squarrosa'** *(right) 6 mm long while leaves of* **'Plumosa'** *are half this size*

Native to Japan, this tree has distinctively small cones and as the type is far less common than its numerous cultivars. These can be divided into those with pendulous foliage such as 'Filifera' and those whose awl-shaped leaves are set at about 45° like 'Plumosa' or approximately at right angles as on 'Squarrosa'.

# Alaska cedar

## *Chamaecyparis nootkatensis*

**Leading shoot** leans

**Crown** is extremely regular

**Cone** 1 cm, globular, with large scale spikes, green with blue bloom, ripens brown over 2 years

**Foliage** very pendulous, in thick, flat, alternate sprays. **Leaves** 2–3 mm, hard, in equal-sized pairs

**30 m. Bark** stringy, peels

One of the parents of Leyland cypress (p 29), Alaska cedar, native from Alaska southward, is recognized by its hanging branchlets and the hooked spines of its cones. Its 'Pendula' form has very pendulous foliage, shorter, upturned branches and 2 cm cones.

# Atlantic white cedar

## *Chamaecyparis thyoides*

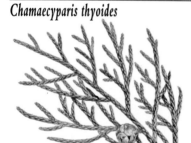

**Foliage** very slender, 1 mm, short, fern-like, angular green or bluish-gray sprays. **Cones** set on small branchlets

**Leaves** often have central resin glands and carry prominent white marks near bases, particularly on undersides

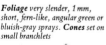

**Leaves** very small, down to 1 mm, acutely pointed, incurved and close-pressed but freer on vigorous growth. **Shoot** becomes brown in 2nd year

**15 m. Crown** columnar or broad conic. **Branches** short. **Bark** stringy. **Cone** 6 mm, glaucous blue-purple ripening brown

This slow-growing white cedar is native to swamps along the eastern American seaboard. Its wood is so durable buried for decades have proved sufficiently strong to use for making roof shingles. Its botanical name comes from a resemblance to Thuja (pp34 –5) and its common name from the paleness of its foliage.

31

# Lawson cypress

## *Chamaecyparis lawsoniana*

**20–35 m**, to 50 m in wild. **Crown** regular in young trees, less so in mature ones. **Stem** often forked. **Foliage** dense, pendulous, becoming spaced in old trees. **Leader** and new shoots always droop. **Terminals** wispy, unbranched near tips

**Cones** 8 mm, globose, ripening from green or blue-green to dark brown. **Scales** (8) have short central spikes

**Bark** smooth, ridged then scaly, dark red-brown, purplish on older trees

♂ **flowers** terminate weakest branchlets

**Foliage** flat, fern-like. **Leaves** grouped in pairs: lateral ones keel-shaped, facing pairs smaller, adpressed. Each leaf has central translucent gland. Between leaf scales, stomata form thin, white lines. Especially clear on foliage underside; give best identification

The Lawson cypress is native to a small area on the Oregon-California border and as a forest species is characterized by a uniform crown of dense, pendent foliage. An expedition sponsored by the Scottish nurseryman Peter Lawson discovered it there in 1854 and the vast array of hardy and easily propagated cultivars raised since then—some 250—have made this "false" cypress one of the most common ornamental conifers. Its varieties can be divided into a group with vivid foliage, one with distinctly shaped habits and a third whose long, awl-like leaves resemble the juvenile leaves of the seedling.

32

# Some cultivars of Lawson cypress

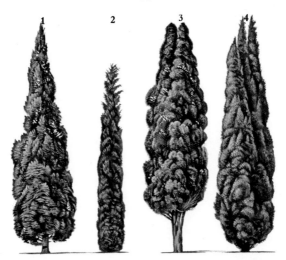

**1 'Allumii'**: 20 m; compact habit, soft foliage. **2 'Columnaris'**: 10 m; dense, consistently narrow crown. **3 'Erecta'**: 25 m; first Lawson cypress cultivar, raised in 1855 from seeds of a Californian type; much-forked crown. **4 'Fletcherii'**: 16 m; juvenile foliage, often on several stems; multiple leaders.

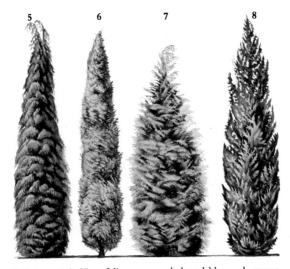

**5 'Intertexta'**: 25 m; foliage sparse, dark and bloomed, sprays pendent. **6 'Lutea'**: 16 m; short, pendulous branchlets; older interior foliage darker. **7 'Stewartii'**: 16 m; ascending branches with sprays decurved below shoot. **8 'Wissellii'**: 25 m; spaced foliage arranged in dense, radiating "spires".

## *Thuja plicata*

**40 m. Crown** conic, broadening with age. **Tip** erect

**Cone** 1 cm, with 10–12 scales, has hooked tip, flask-shaped when closed

**Seed** 5 mm, winged, notched

**Branches** light, even, upswept at ends, can layer in old trees

**Foliage** aromatic

**Bole** fluted

**Bark** dull, stringy

**Foliage** flattened, overlapping. **Leaves** incurved, glossy, paler below with whitish streaks

Also called Giant arborvitae, this majestic tree from the Pacific coast has a light but very even crown and is planted for its light and durable timber. Japanese arborvitae (*T. standishii*) has lighter, blunt, glanded leaves which smell of lemons.

# Chinese arborvitae

## *Biota orientalis*

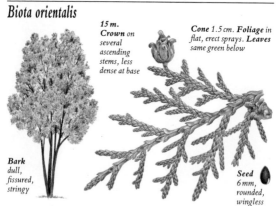

**15 m. Crown** on several ascending stems, less dense at base

**Cone** 1.5 cm. **Foliage** in flat, erect sprays. **Leaves** same green below

**Bark** dull, fissured, stringy

**Seed** 6 mm, rounded, wingless

This species is often included in the Thuja genus and although its foliage bears some resemblance to that of the true thujas, it differs from them in its broader, glaucous cones which have fewer, strongly hooked scales and large, round, wingless seeds.

# Northern white cedar

## *Thuja occidentalis*

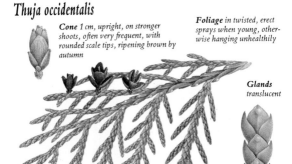

**Cone** *1 cm, upright, on stronger shoots, often very frequent, with rounded scale tips, ripening brown by autumn*

**Foliage** *in twisted, erect sprays when young, otherwise hanging unhealthily*

**Glands** *translucent*

**Foliage** *smells of apples*

**Leaves** *acutely tipped*

**Leaves** *small, 2.5 mm, thick, uniformly paler and glandular below*

Small and slow-growing, Northern white cedar, or Arborvitae, is native from eastern Canada to the Appalachians. Its smooth cones and foliage underside are distinctive. 'Lutea' has a denser crown and stronger branches whose tips bear golden leaves.

# Hiba

## *Thujopsis dolabrata*

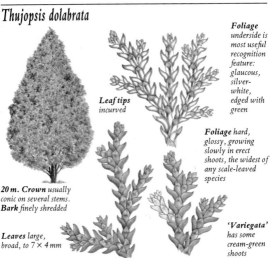

**Foliage** *underside is most useful recognition feature: glaucous, silver-white, edged with green*

**Leaf tips** *incurved*

**Foliage** *hard, glossy, growing slowly in erect shoots, the widest of any scale-leaved species*

**20 m. Crown** *usually conic on several stems.* **Bark** *finely shredded*

**Leaves** *large, broad, to 7 × 4 mm*

**'Variegata'** *has some cream-green shoots*

This Japanese tree rarely grows on a single stem and carries rounded, blue-gray cones. The leaf undersides of Korean arborvitae (*Thuja koraiensis*) are also — sometimes completely — silvery glaucous but its foliage is softer and it has typical *Thuja* cones.

# Incense cedar

## *Calocedrus decurrens*

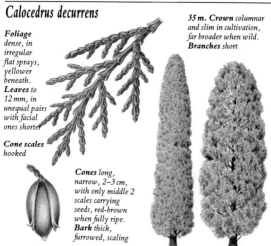

**Foliage** dense, in irregular flat sprays, yellower beneath. **Leaves** to 12 mm, in unequal pairs with facial ones shorter

**Cone scales** hooked

**35 m. Crown** columnar and slim in cultivation, far broader when wild. **Branches** short

**Cones** long, narrow, 2–3 cm, with only middle 2 scales carrying seeds, red-brown when fully ripe. **Bark** thick, furrowed, scaling

A native of Oregon and California, Incense cedar can be confused with the much more common 'Erecta' cultivar of Lawson cypress (p 33) but is distinguished by its irregular foliage, short, upswept branches and the yellower undersides of its leaves. Its fragrant timber gives the tree its name.

# Juniper

## *Juniperus communis*

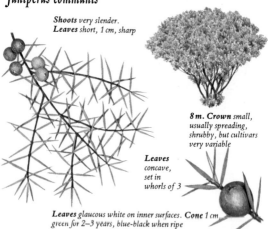

**Shoots** very slender. **Leaves** short, 1 cm, sharp

**8 m. Crown** small, usually spreading, shrubby, but cultivars very variable

**Leaves** concave, set in whorls of 3

**Leaves** glaucous white on inner surfaces. **Cone** 1 cm green for 2–3 years, blue-black when ripe

Juniper will grow on both acid and alkaline sites and has a very wide distribution throughout the northern hemisphere. The bright green leaves of Temple juniper (*J. rigida*) are softer and longer while Alerce (*Fitzroya cuppressoides*) has spreading blue-green leaves with two silver bands on each side.

# Eastern red cedar

## *Juniperus virginiana*

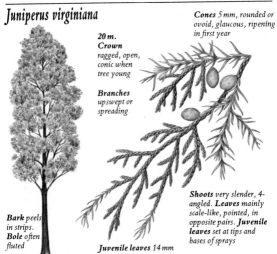

**Cones** *5 mm, rounded or ovoid, glaucous, ripening in first year*

**20 m. Crown** *ragged, open, conic when tree young*

**Branches** *upswept or spreading*

**Shoots** *very slender, 4-angled.* **Leaves** *mainly scale-like, pointed, in opposite pairs.* **Juvenile leaves** *set at tips and bases of sprays*

**Bark** *peels in strips.* **Bole** *often fluted*

**Juvenile leaves** *14 mm*

Eastern red cedar, or Pencil cedar, native to the eastern USA, is the tallest of the junipers and provides useful timber, especially for pencils. Alligator juniper (*J. deppeana*), from the southwest USA, has distinctive bark, up to 10 cm thick, deeply fissured into 5 cm squares like an alligator's skin. Its foliage is blue-green.

# Chinese juniper

## *Juniperus chinensis*

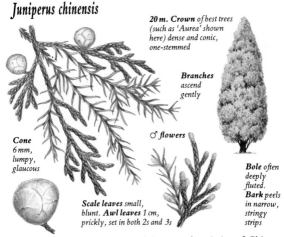

**20 m. Crown** *of best trees (such as 'Aurea' shown here) dense and conic, one-stemmed*

**Branches** *ascend gently*

**Cone** *6 mm, lumpy, glaucous*

♂ **flowers**

**Bole** *often deeply fluted.* **Bark** *peels in narrow, stringy strips*

**Scale leaves** *small, blunt.* **Awl leaves** *1 cm, prickly, set in both 2s and 3s*

'Aurea' is the most popular of the named varieties of Chinese juniper which have been developed, many of them shrubs, and produces masses of flowers. Western juniper (*J. occidentalis*) has glandular and scaly bark while Rocky mountain juniper (*J. scopulorum*) has glandular leaves, often glaucous.

# Redwood family Taxodiaceae

This ancient group of large, beautiful trees comprises 10 genera from North America, eastern Asia and Tasmania. Its members usually have evergreen leaves, which are flat, linear or awl-shaped and set spirally along the shoot, and globular woody cones, whose peltate scales are also arranged spirally. All redwoods are monoecious. They have thick barks that are red-brown and fibrous. Some genera are deciduous.

## Giant sequoia

### *Sequoiadendron giganteum*

**Leaves** small, to 7 mm, awl-shaped, hard, dotted with white stomata

**Foliage** blue-gray when young then shiny, dark green on older shoots

**Shoot** stout, initially covered by leaves, later developing gray-brown fissures

**Cone** large, 6 cm, 35–40 scales, green for 2 years, then brown

**Branches** have tips upswept

**50 m. Crown** often has rounded apex. **Bole** tapers, fluted. **Bark** often dimpled by Tree-creepers using it for winter roosting

**Bark** very thick, to 30 cm, soft and resilient, can withstand punching

Giant sequoia is the world's largest (though not the tallest) living phenomenon and is native to 72 groves on the high, western slopes of the Californian Sierra Nevada where it was first discovered in 1841. The largest individual there, named "General Sherman", is 83 m tall, has a trunk diameter of 10 m and weighs 1,000 tonnes, yet must have developed over the centuries from a seed weighing a mere 5 milligrammes. Such specimens may live for 4,000 years although the average age of native species is only a quarter of this. In Europe, growth rates have exceeded 50 m in a century. Giant sequoia have deep roots to withstand long, dry summers and their thick bark gives protection against forest fires.

# California redwood

## *Sequoia sempervirens*

**Shoot** *green.* **Strong shoot leaves** *small, 6–8 mm, scale-like, awl-shaped, adpressed around stem with tips incurved*

**Side-shoot leaves** *1–2 cm, tapering to hard apexes with 2 white bands below, set in flat sprays*

**Cone** *3 cm.* **Scales** *(15–20) ripen in first season*

**50 m. Crown** *thin, columnar conic, becoming flat-topped.* **Bole** *straight, tapering*

**Bark** *thick, very soft, fibrous; bright red-brown on young trees, darker and fissured on old ones*

This redwood is the tallest tree in the world, reaching 112 m (the height of St Paul's Cathedral, London) in its native California where it thrives in the damp atmosphere of the narrow, coastal "fogbelt". Chinese fir (*Cunninghamia lanceolata*) has a similar bark but much longer, more broadly based leaves.

# Bald cypress

## *Taxodium distichum*

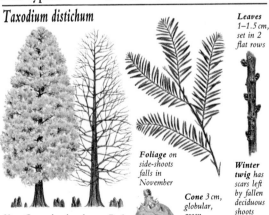

**Leaves** *1–1.5 cm, set in 2 flat rows*

**Foliage** *on side-shoots falls in November*

**Winter twig** *has scars left by fallen deciduous shoots*

**Cone** *3 cm, globular, green, ripening brown in first season*

**30 m. Crown** *has domed apex.* **Bark** *fibrous.* **Aerial roots** *(pneumatophores) supply oxygen to trees growing in water-logged sites*

Sometimes confused with *Metasequoia* (p 40) from which it can be distinguished by its alternate foliage, this tree from the southern USA prefers waterside sites but grows well on any fairly moist soil. The closely related Pond cypress (*T. ascendens*) has erect, spiky shoots of shorter, radially arranged leaves.

# Dawn redwood

## *Metasequoia glyptostroboides*

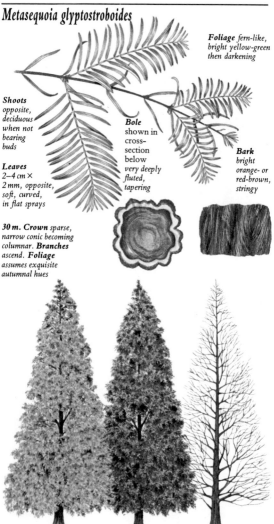

**Foliage** *fern-like, bright yellow-green then darkening*

**Shoots** *opposite, deciduous when not bearing buds*

**Bole** *shown in cross-section below very deeply fluted, tapering*

**Bark** *bright orange- or red-brown, stringy*

**Leaves** *2–4 cm × 2 mm, opposite, soft, curved, in flat sprays*

**30 m. Crown** *sparse, narrow conic becoming columnar.* **Branches** *ascend.* **Foliage** *assumes exquisite autumnal hues*

Long thought to be extinct, this fine tree was located in south-eastern China as recently as 1941 since when its hardiness, the ease with which it propagates and a growth rate which can average a meter per annum have made it a popular ornamental. Its botanical name indicates affinities with *Sequoia* (p 39) and the rare, deciduous Chinese swamp cypress (*Glyptostrobus lineatus*) but it is more easily confused with Bald cypress (p 39), whose similar foliage is also deciduous. The leaves and lateral shoots of Dawn redwood, however, are longer and *opposite*. It is unique in carrying its side-buds *below* its shoots and not in their axils.

# Japanese cedar

## *Cryptomeria japonica*

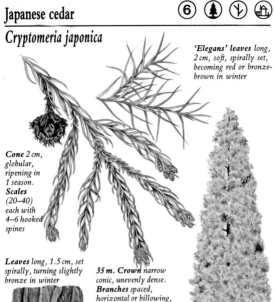

*'Elegans' leaves* long, 2 cm, soft, spirally set, becoming red or bronze-brown in winter

*Cone* 2 cm, globular, ripening in 1 season. *Scales* (20–40) each with 4–6 hooked spines

*Leaves* long, 1.5 cm, set spirally, turning slightly bronze in winter

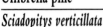

*35 m. Crown* narrow conic, unevenly dense. *Branches* spaced, horizontal or billowing, tips upswept. *Bark* soft, fibrous, fissured, peeling on old trees

As an introduced ornamental, this tree never reaches the heights it attains in its native China and Japan. The Tasmanian *Athrotaxis* species have similar cones but shorter leaves: King William pine (*A. selaginoides*) has hard, shiny leaves of 1 cm, Summit cedar (*A. laxifolia*) awl-like ones to 2 mm and Tasmanian cedar (*A. cupressoides*) adpressed scale leaves.

# Umbrella pine

## *Sciadopitys verticillata*

*Leaves* very long, 10–15 cm, fused in grooved pairs which glossy above and bright yellow below

*20 m. Crown* narrow to broad conic. *Branches* short, upturned and spreading. *Bark* dark red-brown, peeling in strips

*Foliage* set round shoots in whorls

Native to the mountains of Japan, where it is an important timber tree, this ornamental species grows very slowly in parks and large gardens. Its distinctive foliage of long and whorled, glossy "double" leaves distinguishes it from all other conifers and offers immediate identification.

# Pine family Pinaceae

The Pine family is the most varied of all those groups of trees which bear cones. Besides the genus *Pinus* itself, the family also contains those of *Abies*, *Picea*, *Tsuga*, *Pseudotsuga* and *Larix* as well as *Pseudolarix*, *Cathaya* and *Keteleeria* (these latter three are fairly obscure Chinese genera and, because they are so seldom found elsewhere, not included in this book.)

Together, all the Pinaceae comprise well over 250 species, all of which are native to the northern hemisphere. They all have woody cones with spirally arranged scales and linear flat leaves (usually called needles) which are attached to the shoot in a variety of different ways.

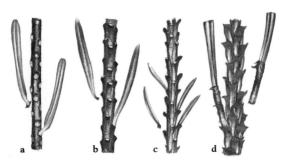

a   b   c   d

The illustrations above demonstrate these differences. The leaves of the firs (*Abies*) leave a depressed or concave scar (**a**) while a slightly raised scar, together with a minutely stalked leaf (**b**) shows the tree is almost certainly a Douglas fir (*Pseudotsuga*). Spruces (*Picea*) are immediately notable for the prominent woody peg (*pulvinus*) that is left where the leaf (**c**) is pulled away or falls from it naturally. The leaves of pines (*Pinus*) are set in fascicles of two, three or five and bound by a basal sheath by which they are attached to the shoot (**d**). These fascicles are extensions of short or spur shoots and—despite the number of their leaves—can always be brought together to make up a simple, but divided cylinder. This adaptation probably allows the leaves to maximize their surface area for the purposes of transpiration, an essential part of photosynthesis.

The length of the shoot can also be a guide to genus; while all the family produce long shoots, some genera, namely the pines, larches (*Larix*) and cedars (*Cedrus*) also produce much shorter shoots. These grow from a bud in the axil of a needle on a long shoot, and eventually a new whorl of needles is formed. The bud in the center of this new whorl may either be the starting point for subsequent growth or may remain dormant for years. The shorter shoots of the larches grow for several years and have many deciduous needles, a feature unique in this family.

The majority of trees in Pinaceae grow pendent cones but *Abies*, *Cedrus* and *Larix* have cones which remain erect after the female flowers have been fertilized. Those of larches ripen over one season; those of the firs and cedars usually take longer and are further distinguished by their cones being deciduous, falling to the ground after they have opened to release their seeds. The central core of these cones, a long spike or "candle", is left standing on the shoot.

.All other genera have pendent cones whose scales remain attached to their central axis and eventually fall to the ground intact, having released their seeds.

# Firs *Abies*

Most firs have short leaves set in flat ranks which become more assurgent in the upper crown; silver firs are so called because of the whitish underside of their foliage. Their deciduous cones develop at the top of the tree. About 40 species exist.

## Silver fir

### *Abies alba*

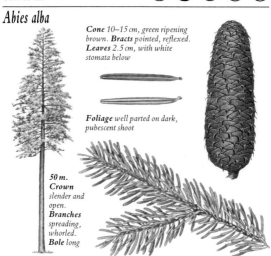

*Cone 10–15 cm, green ripening brown. Bracts pointed, reflexed. Leaves 2.5 cm, with white stomata below*

*Foliage well parted on dark, pubescent shoot*

*50 m. Crown slender and open. Branches spreading, whorled. Bole long*

Silver fir is an important forest tree in its native area, the mountains of central Europe. It has thick leaves with notched tips, and non-resinous buds. King Boris fir (*A. borisii-regis*) has denser foliage, narrower leaves to 3 cm and pale-haired twigs.

## Caucasian fir

   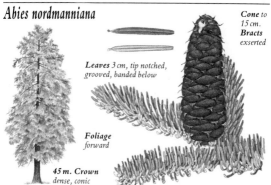 

### *Abies nordmanniana*

*Cone to 15 cm. Bracts exserted*

*Leaves 3 cm, tip notched, grooved, banded below*

*Foliage forward*

*45 m. Crown dense, conic*

This fir has a more luxuriant crown than *A. alba* and its forward-pointing foliage persists for about 6–8 years. Bornmüller fir (*A. bornmuellerana*) has longer leaves with spots of stomata by the upperside tips, shiny red-brown shoots and sticky buds.

# Spanish fir

## *Abies pinsapo*

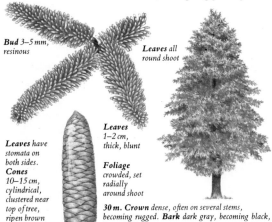

*Shoot brownish, becoming orangey in 2nd year*

**Bud** 3–5 mm, resinous

**Leaves** all round shoot

**Leaves** have stomata on both sides. **Cones** 10–15 cm, cylindrical, clustered near top of tree, ripen brown

**Leaves** 1–2 cm, thick, blunt

**Foliage** crowded, set radially around shoot

**30 m. Crown** dense, often on several stems, becoming rugged. **Bark** dark gray, becoming black, in small plates

Spanish fir only occurs wild in southern Spain but is fairly widely planted in Europe. It is also called Hedgehog fir because of its stiff foliage; that of Algerian fir (*A. numidica*) is parted below and has a prominent band of stomata at its leaf tips.

# Greek fir

## *Abies cephalonica*

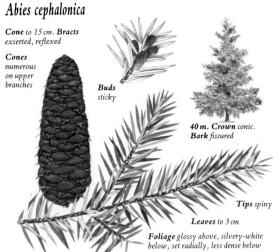

**Cone** to 15 cm. **Bracts** exserted, reflexed

**Cones** numerous on upper branches

**Buds** sticky

**40 m. Crown** conic. **Bark** fissured

**Tips** spiny

**Leaves** to 3 cm

**Foliage** glossy above, silvery-white below, set radially, less dense below

With a massive bole and heavy branches which sometimes rise as secondary leaders, Greek fir is often the bulkiest of the firs. *A. cephalonica* var. *apollonis* has denser, blunter, forward-pointing leaves, mostly arranged above the shoot.

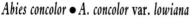

## Abies concolor • A. concolor var. lowiana

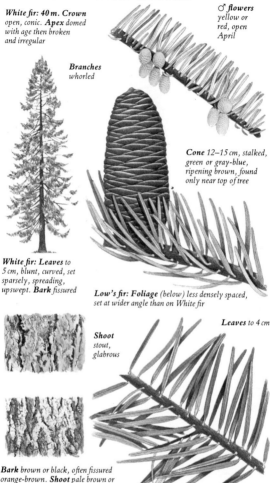

**White fir: 40 m.** Crown open, conic. **Apex** domed with age then broken and irregular

♂ **flowers** yellow or red, open April

**Branches** whorled

**Cone** 12–15 cm, stalked, green or gray-blue, ripening brown, found only near top of tree

**White fir: Leaves** to 5 cm, blunt, curved, set sparsely, spreading, upswept. **Bark** fissured

**Low's fir: Foliage** (below) less densely spaced, set at wider angle than on White fir

**Leaves** to 4 cm

**Shoot** stout, glabrous

**Bark** brown or black, often fissured orange-brown. **Shoot** pale brown or bright green, turning copper-brown or orange-brown in 2nd year

White fir grows wild in the western USA and parts of Mexico, and is recognizable by its long, assurgent, bluish leaves which smell of lemons when crushed. Low's fir, from Oregon and California, has longer leaves which either spread flat along the shoot or rise on both sides to form a wide U-shaped "groove". While it therefore combines several features of White fir, which grows to the south of its range, with several of Grand fir, which is native further north, Low's fir is ultimately distinguishable from other firs by the unique combination of lax, bluish leaves with a fissured bark usually resembling that of Douglas fir (p 61).

# Grand fir

## *Abies grandis*

**Foliage** *flat on shoots.* **Leaves** *5 cm.* **Buds** *2 mm, conic, gray-white, and resinous*

*60 m.* **Crown** *conic*

**Cone** *9 cm, tapering resinous, ripening brown*

**Branches** *whorled, usually level.* **Bark** *smooth, shiny when young, later cracked*

A very fast-growing species, Grand fir is an important forest tree in western North America, identified by its flat foliage which becomes more assurgent in the upper crown. West Himalayan fir (*A. pindrow*) has large globular buds and longer leaves to 9 cm, spreading down at the sides of its glabrous, ash-gray shoots.

# Pacific fir

## *Abies amabilis*

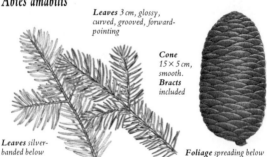

**Leaves** *3 cm, glossy, curved, grooved, forward-pointing*

**Cone** *15 × 5 cm, smooth.* **Bracts** *included*

**Leaves** *silver-banded below*

**Foliage** *spreading below*

The shapely crown and rich foliage of this fir justify its specific name which may be translated as "lovely". It is native from British Columbia to California. Maries fir (*A. mariesii*) is its closest relative and its 2 cm leaves are glossier. Its shoot has dense orange-red, not light brown, pubescence.

# Noble fir

## *Abies procera*

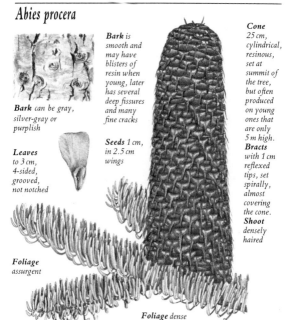

**Bark** is smooth and may have blisters of resin when young, later has several deep fissures and many fine cracks

**Bark** can be gray, silver-gray or purplish

**Leaves** to 3 cm, 4-sided, grooved, not notched

**Seeds** 1 cm, in 2.5 cm wings

**Cone** 25 cm, cylindrical, resinous, set at summit of the tree, but often produced on young ones that are only 5 m high.

**Bracts** with 1 cm reflexed tips, set spirally, almost covering the cone.

**Shoot** densely haired

**Foliage** assurgent

**Foliage** dense

Noble fir is remarkable for the size of its cones which may contain up to 1,000 seeds and for its silvery fissured mature bark. Its lower branches carry many crimson male flowers in spring.

# Red fir

## *Abies magnifica*

**40 m.**
**Crown** narrow, columnar-conic and symmetrical.
**Branches** short and whorled, assurgent

**Bark** thick and corky.
**Bole** stout and long, tapering

**Cone** to 20 × 10 cm, smooth, initially golden green, ripening purple-brown.
**Bracts** included

**Leaves** 4 cm, lax, curved, 4-sided, ungrooved

This fir is native to Oregon and California and while closely related to Noble fir, has ungrooved, longer and less densely set foliage. Its name comes from the red bark of mature trees. The cones of Shasta fir (*A. magnifica* var. *shastensis*) have exserted bracts.

# Veitch fir

## Abies veitchii

**Cone** 8 cm, cylindrical, flat-topped, purple-blue initially, ripens brown, smoothly wrinkled. **Bracts** slightly exserted

**Foliage** all pointing forwards, spreading below, rising at 45° to shoot above

**Bud** 3 mm, shiny, red-purple

**Bark** dark gray-green, with white patches on old trees

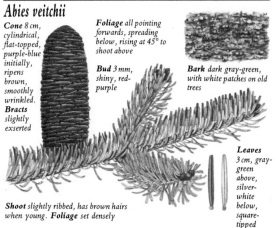

**Leaves** 3 cm, gray-green above, silver-white below, square-tipped

**Shoot** slightly ribbed, has brown hairs when young. **Foliage** set densely

Veitch firs form trees to 20 m with tapered, flat-topped crowns. On trees whose crowns reach the ground the lowest branches are very upswept revealing the silver underside of their foliage. It is a native of Japan, as is Sakhalin fir (*A. sachalinensis*), with longer, narrower, bright green leaves to 3.5 cm.

# Korean fir

## Abies koreana

**15 m. Crown** conic with slightly ascending branches. **Leaves** spreading below shoot, curving upwards above

**Leaves** 1–1.5 cm, glossy green or yellow-green above, tips often white

**Bark** shiny, dark brown to black, smooth, spotted conspicuously with lenticels. **Leaves** vividly white underneath, rather radial and spaced along shoot

**Shoot** fawn. **Buds** small, globular, initially pale brown but soon covered with white resin

**Cone** 7 cm, pointed, purple, ripening brown. **Bracts** reflexed, very exserted

This silver fir, first discovered in 1907 on an island off Korea, usually only manages to grow to 10 m in 40 years. It is very free in producing its small violet cones, often as a young tree less than 1 m tall. The strongly exserted and reflexed bracts show clearly how cone scales in the Pine family are radially arranged.

# Balsam fir

## *Abies balsamea*

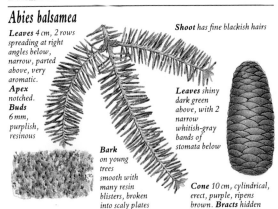

**Leaves** 4 cm, 2 rows spreading at right angles below, narrow, parted above, very aromatic.
**Apex** notched.
**Buds** 6 mm, purplish, resinous

**Shoot** *has fine blackish hairs*

**Leaves** shiny dark green above, with 2 narrow whitish-gray bands of stomata below

**Bark** on young trees smooth with many resin blisters, broken into scaly plates

**Cone** 10 cm, cylindrical, erect, purple, ripens brown. **Bracts** hidden

Often used as a Christmas tree, Balsam fir, from eastern Canada and northern New England, reaches 15 m, with a spire-like crown. The cones of the closely related Fraser fir (*A. fraseri*), from the Appalachians, have strongly exserted and reflexed bracts.

# Subalpine fir

## *Abies lasiocarpa*

**Leaves** 4.5 cm × 1 mm, flat, dense, spreading below, pointing forward above and at sides, gray-green, white bands on both sides

**Shoot** stout, with dense rufous hairs, later gray-brown, shiny

**Buds** 5 mm, resinous

**30 m. Crown** narrow spire, dense, furnished with branches to the ground, lower ones pendulous, the rest horizontal.
**Bark** smooth, gray, with resin blisters, later fissured, scaly

**Cone** 10 cm, cylindrical, erect, numerous on topmost branches, dark purple-gray at first, ripening brown, soon disintegrating.
**Bracts** included.
**Scales** hairy

Subalpine fir is often the last tree before the treeline in its native Rockies, and at higher altitudes has a less regular habit. In the southern part of its range it becomes Arizona or Cork fir (*A. lasiocarpa* var. *arizonica*), with bluish foliage and corky bark.

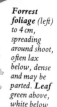
## *Abies delavayi* var. *forrestii* • *A. delavayi* var. *georgei*

**Forrest fir cone** (right) is 7–12 cm long, cylindrical or barrel-shaped, top dimpled, violet, ripening brown over winter. **Bracts** exserted, often reflexed with very prominent awl-like cusps to 5 mm.

**George fir cone** (right) often larger, occasionally to 15 cm. **Bracts** exserted, with cusps up to 1 cm, pointing upright except near base of cone. **Bract edges** well exposed, bright blue-purple with light brown edges

**Forrest fir shoot** (below) stout, red-brown, usually glabrous and finely roughened; in second year deeper color with pale fissures. **Foliage** may be radial

**George fir shoot** (above) brownish orange, with a dense, short pubescence of the same color. **Leaves** close more over top of shoot. **Foliage** short, perpendicular

**Forrest foliage** (left) to 4 cm, spreading around shoot, often lax below, dense and may be parted. **Leaf** green above, white below

**George fir leaves** (left) shorter, to 2.5 cm, off-white below, gray bloomed above

These two firs were discovered in China and introduced into gardens by George Forrest, after whom they are named. George fir is chiefly distinguished by the densely pubescent shoots and longer cusps, and although both trees grow to 25 m, George fir has a more columnar and denser crown than Forrest fir. Related species from the Himalayas and west China include Himalayan fir (*A. spectabilis*), which has ash-gray or light brown shoots pubescent in deep grooves and Delavay fir (*A. delavayi*), which has bright violet narrow cones, maroon shoots, orange buds, and inrolled leaf margins which make the leaves narrow and square-tipped. Farges fir (*A. fargesii*) has stout 2.5 cm needles on glossy, purple shoots and conic, purple buds.

# Nikko fir

## *Abies homolepis*

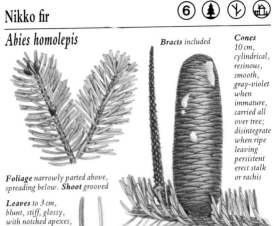

*Bracts* included

**Cones**
10 cm,
cylindrical,
resinous,
smooth,
gray-violet
when
immature,
carried all
over tree;
disintegrate
when ripe
leaving
persistent
erect stalk
or rachis

**Foliage** *narrowly parted above, spreading below.* **Shoot** *grooved*

**Leaves** *to 3 cm, blunt, stiff, glossy, with notched apexes, grooved, with 2 silvery bands below*

Nikko fir, tolerant of urban pollution, has strongly ridged and grooved glabrous shoots. Min fir (*A. recurvata*) has smooth shoots, more ovoid, 8 cm cones, and bluntly pointed needles, green on both surfaces, that may point backwards.

# Santa Lucia fir

## *Abies bracteata*

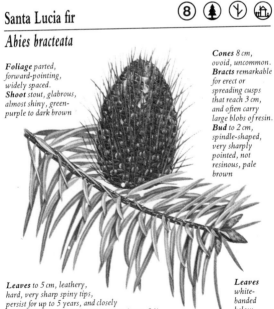

**Foliage** *parted, forward-pointing, widely spaced.* **Shoot** *stout, glabrous, almost shiny, green-purple to dark brown*

**Cones** *8 cm, ovoid, uncommon.* **Bracts** *remarkable for erect or spreading cusps that reach 3 cm, and often carry large blobs of resin.* **Bud** *to 2 cm, spindle-shaped, very sharply pointed, not resinous, pale brown*

**Leaves** *to 5 cm, leathery, hard, very sharp spiny tips, persist for up to 5 years, and closely resemble those of California nutmeg (see p 24)*

**Leaves** *white-banded below*

Unique in its cones, its beech-like buds and its foliage, this rare fir grows wild in the Santa Lucia mountains of California. Manchurian fir (*A. holophylla*) has similar assurgent foliage but ovoid-conic, resinous buds.

51

# True Cedars *Cedrus*

True cedars develop two types of foliage and have deciduous cones which ripen over two years and disintegrate *in situ* to release triangular-winged seeds. "Atlas—ascending, Deodar—drooping, Lebanon—level" can be useful mnemonic for identifying cedars by their *growing* branch tips.

## Cedar of Lebanon ⑦ 🌲 ✳ 🏠

### *Cedrus libani*

**40 m. Crown** broad, flat-topped. Large, horizontal layers of dense foliage

**Bark** gray-brown, fissured; smooth, dark gray on young trees

**Branches** massive, arching then level. **Bole** huge

**Foliage** dense, darkest green of cedars, can be gray or glaucous. **Buds** 2–3 mm, brown, ovoid. **Shoots** brown, pubescent

**Cone** from 8–14 cm. Apex sometimes depressed. Resinous

**Leaves** narrow. Taper to sharp, translucent tips. On young, long shoots: 2 cm, set singly and spirally; on older 2–3 mm spur-shoots: 3 cm, in whorls of 10–20

Cedar of Lebanon grows naturally in Asia Minor and has become a familiar ornamental in parks in the south. Cyprus cedar (*C. brevifolia*) is less common and has shorter leaves, narrower, more ovoid cones and a more conical crown.

# Atlas cedar

## *Cedrus atlantica*

**Cone** 8 cm, usually depressed

♂ *flowers reach 5 cm by Sept when pollen is shed and dispersed by wind*

**40 m. Crown** *broad, conic.* **Branches** *widely spaced.* **Leaves** *2.5 cm on long shoots; 2 cm, in whorls of 30–45 on 1–2 cm spur-shoots.* **Buds** *ovoid, 2–3 cm*

Native to the mountains of north Africa, wild Atlas cedars are found in both green and glaucous forms. The very blue 'Glauca' clone shown here is the one most commonly encountered as an ornamental and derives its color from the wax coating of its leaves.

# Deodar cedar

## *Cedrus deodara*

**40 m. Crown** *broad, columnar, conic.* **Leader** *and long shoots always droop.* **Stem** *single. Young trees blue-gray*

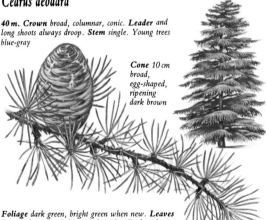

**Cone** 10 cm broad, egg-shaped, ripening dark brown

**Foliage** *dark green, bright green when new.* **Leaves** *5 cm on long shs, 3.5 cm on short shs.* **Buds** *only 1 mm*

In the western Himalayas where they grow wild, Deodars can reach 70 m. Pendulous branchlets on spreading and slightly downswept branches are their most distinctive features.

# Larch *Larix*

Some leaves of these deciduous conifers are set singly on long shoots, but most foliage is set in whorls on short spur-like shoots. The small, erect, persistent cones and the short and long shoots are the key identifying features.

## European larch • Hybrid larch

## *Larix decidua* • *Larix eurolepis*

**Shoot** straw yellow, glabrous. **Cones** to 4 cm, narrow, blunt erect scales

**Leaves** 3 cm, set in whorls of 20–30, blunt

**Leaves** not banded below, on short 1 mm pegs

♀ **flowers** 1 cm, erect.
♂ **flowers** pendent, whitish

**Crown** conic

45 m. **Crown** open

**Hybrid shoot** (below) orange-brown

**Hybrid cones** (above) have scales more reflexed

**Hybrid leaves** to 5 cm

**Bark** smooth initially, later scaly, ridged, fissured, dark pink

**Foliage** pendulous. **Bole** very upright

**Lower branches** more drooping. **Shoots** long, pendulous

Native to the mountains of northern and central Europe, European larch is the only European conifer to shed all its leaves annually. Hybrid, or Dunkeld, larch is a natural cross between European and Japanese larch, identifiable by shoots and cones. Western larch (*L. occidentalis*), from the northern Rockies, differs in its stout shoots and 5 cm cones with long exserted bracts.

# Japanese larch

## *Larix kaempferi*

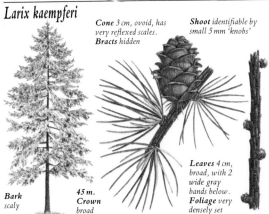

**Cone** 3 cm, ovoid, has very reflexed scales. **Bracts** hidden

**Shoot** identifiable by small 5 mm 'knobs'

**Leaves** 4 cm, broad, with 2 wide gray bands below. **Foliage** very densely set

**Bark** scaly

**45 m. Crown** broad

This species, native to Mount Fuji, is more vigorous than European larch and forms a shorter, stouter tree with heavier branches. It can be distinguished by its purplish-red shoots, wider leaves and squatter cones with scales reflexed like rose petals.

# Tamarack

## *Larix laricina*

**20 m. Crown** slender

**Cones** 2 cm, ovoid, with few rounded scales incurved at tips and very erect

**Leaves** at tips set singly

**Leaves** 2.5 cm, 12–20 set on each spur-shoot

**Shoot** pinkish initially, then pale brown. **Buds** red-brown, resinous

**Branches** often twist

**Bole** long, straight
**Bark** pinkish-brown

**Leaves** have 2 pale bands on both sides when new, later plain dark green above

Tamarack is the most widely distributed conifer in North America growing across Canada from Alaska to the Atlantic, and as far south as Pennsylvania in the USA, in anything from swamps to sub-Arctic conditions. Dahurian larch (*L. gmelinii*) has a gaunt crown and smaller leaves.

# Spruces *Picea*

Spruces have single pointed needles which are set on a *pulvinus*, an extension of the shoot. When the needles fall they leave behind a short peg, making the bare shoots prickly.

## Norway spruce

### *Picea abies*

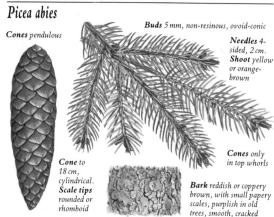

**Cones** *pendulous*

**Buds** *5 mm, non-resinous, ovoid-conic*

**Needles** *4-sided, 2 cm.* **Shoot** *yellow or orange-brown*

**Cone** *to 18 cm, cylindrical.* **Scale tips** *rounded or rhomboid*

**Cones** *only in top whorls*

**Bark** *reddish or coppery brown, with small papery scales, purplish in old trees, smooth, cracked*

Norway spruce, a familiar "Christmas tree", native to central and northern Europe, has longer needles and longer, less woody cones than Siberian spruce (*P. obovata*), from northern Eurasia, and Wilson spruce (*P. wilsonii*) which also has ash gray shoots.

## Caucasian spruce

### *Picea orientalis*

**Needles** *very short, 8 mm, squarish, rounded at tips, shiny dark green, parted below*

**Shoots** *of young trees long, twisted; in old trees straight, projecting from crown*

**35 m. Crown** *narrow conic*

**Shoot** *whitish-brown, densely hairy.* **Buds** *4 mm, with red tips to basal scales*

**Crown** *open when young, dense when old.* **Bark** *gray, smooth, pinkish with age*

Caucasian spruce is unique in its short needles, and has spindle-shaped, often curved cones that grow to 7 cm. Likiang spruce (*P. likiangensis*) has buff shoots, blue-gray needles and papery purple cones to 15 cm.

# Serbian spruce

## *Picea omorika*

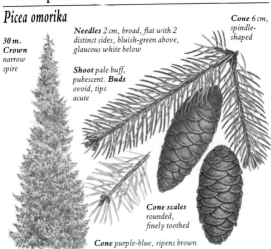

**30 m. Crown** narrow spire

**Needles** *2 cm, broad, flat with 2 distinct sides, bluish-green above, glaucous white below*

**Cone** *6 cm, spindle-shaped*

**Shoot** *pale buff, pubescent.* **Buds** *ovoid, tips acute*

**Cone scales** *rounded, finely toothed*

**Cone** *purple-blue, ripens brown*

Native to a single Yugoslavian river valley but widely planted throughout Europe, this spruce owes its spire-like habit not to the branches being short, but to their lying recumbent down the stem before curving out, an adaptation to prevent snow damage.

# Sitka spruce

## *Picea sitchensis*

**Shoot** *glabrous, whitish, grooved.* **Buds** *ovoid, slightly resinous purple*

**Cones** *8 cm, cylindrical.* **Scales** *papery thin, toothed, ripen yellow-brown.* **Bark** *purple-gray, coarse flaking plates in old trees*

**Leaves** *3 cm, bright green with 2 narrow lines above, 2 blue-gray bands below; imbricate above*

An important timber tree, Sitka spruce is native along the west coast of North America, where it can grow to 80 m, making it the tallest of the spruces. Jezo spruce (*P. jezoensis*), from north-eastern Asia, has denser, assurgent blunt leathery needles to 1.5 cm, and a gaunt habit. Sargent spruce (*P. brachytyla*), from western China, has leaves to 1.5 cm which are silvery white below and curve down at the sides. Its 13 cm cones are conical.

57

# White spruce

## *Picea glauca*

**Crushed foliage** *very odorous*

**Buds**
*5 mm,
ovoid,
smooth,
shiny pale
brown*

**Cone** *6 cm,
ovoid.* **Scales**
*smooth,
round, thin
but woody*

**Foliage** *assurgent, all
above shoot.* **Needles**
*even length, 1.5 cm,
round, gray or blue-gray
with narrow white bands*

**Shoot** *glabrous, white,
later orange-brown*

White spruce, which has a wide range across Canada and northern USA, grows to 20 m with a narrow conic crown. Engelmann spruce (*P. engelmannii*) differs in its longer, softer, 2.5 cm needles, hairy shoots and papery scales. Native to the eastern Rockies, it has a red-brown, not gray, bark.

# Colorado blue spruce

## *Picea pungens* 'Glauca'

**Needles** *2 cm, arranged
radially but upswept below;
4-sided, stiff, sharp.*
**Buds** *1 cm, with long
slender scales at base*

**25 m.
Crown**
*columnar-
conic,
dense*

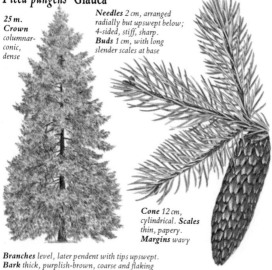

**Cone** *12 cm,
cylindrical.* **Scales**
*thin, papery.*
**Margins** *wavy*

**Branches** *level, later pendent with tips upswept.*
**Bark** *thick, purplish-brown, coarse and flaking*

This form of the normally blue-green foliaged Colorado spruce is the one usually cultivated. Dragon spruce (*P. asperata*) has gray exfoliating papery bark and lacks a ring of scales at the base of the bud; its cones have round woody scales. Tigertail spruce (*P. polita*) has viciously sharp, shiny dark green radial leaves and cones with rounded scales.

58

# Black spruce

## *Picea mariana*

**20 m. Crown** *conic, sometimes very narrow, appears dark blue-gray from distance.* **Branches** *horizontal or slightly pendent, reach to ground; when pressed there by snow for long periods they frequently layer*

**Leaves** *1.5 cm, dark blue-green with 2 bluish-white bands below; spreading around shoot, looser below and pressed down above*

**Leaves** *stiff, four-sided*

**Shoot** *densely hairy*

**Bark** *gray, flaky.* **Bud** *ovoid, hairy, red-brown.* **Cone** *persistent, clustered in crown*

**Cone** *3.5 cm, ovoid, pendent, often curved, grow profusely even on young trees*

Black spruce, so called because of the dark appearance of its foliage from afar, grows across Canada and in New England and is often the last tree before the tundra in the north. It frequents wet sites and because of layering saplings may occur in rings.

# Red spruce

## *Picea rubens*

**30 m. Crown** *narrowly conic* **Branches** *at first ascend, later curve down and out*

**Leaves** *1.5 cm, 4-sided, yellow-green, spreading then curving inward*

**Cone** *5 cm, ovoid-oblong, reddish, soon falling*

**Bark** *red- or gray-brown*

Red spruce is related to Black spruce but grows only in southeastern Canada and New England, and is distinguishable by the redness of its cones and its inner bark, which unlike the outer bark is always red-brown. Its ovoid, acute buds are hairy.

59

# Brewer spruce

## *Picea breweriana*

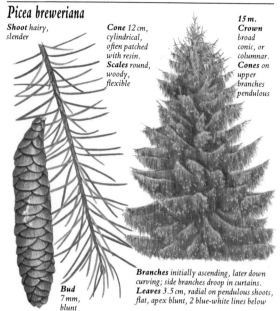

**Shoot** hairy, slender

**Cone** 12 cm, cylindrical, often patched with resin. **Scales** round, woody, flexible

**15 m. Crown** broad conic, or columnar. **Cones** on upper branches pendulous

**Branches** initially ascending, later down curving; side branches droop in curtains. **Leaves** 3.5 cm, radial on pendulous shoots, flat, apex blunt, 2 blue-white lines below

**Bud** 7 mm, blunt

A native of Oregon and California, this fine weeping tree has branchlets drooping vertically either side of the main branches. Sikkim spruce (*P. spinulosa*) has white glabrous shoots on an open, less pendent crown, and its leaves are not as flat.

# Morinda spruce

## *Picea smithiana*

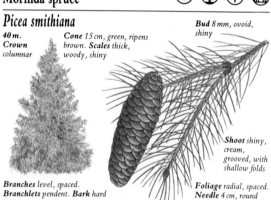

**40 m. Crown** columnar

**Cone** 15 cm, green, ripens brown. **Scales** thick, woody, shiny

**Bud** 8 mm, ovoid, shiny

**Shoot** shiny, cream, grooved, with shallow folds

**Branches** level, spaced. **Branchlets** pendent. **Bark** hard

**Foliage** radial, spaced. **Needle** 4 cm, round

Morinda spruce is native from Afghanistan to Nepal, and has the largest cones of the genus, up to 20 × 5 cm, borne throughout the upper crown. Schrenk spruce (*P. schrenkiana*) is similar, but has less weeping branchlets, gray-green leaves less radially arranged and long pale buds.

# Douglas fir

## Pseudotsuga menziesii

**Douglas foliage** densely set, parted above (left), spreads below, emits powerful sweet aroma, can be any shade of green

**60 m. Crown** columnar

**Leaves** to 2.5 cm, with bands below

**Buds** 7 mm, conic, not resinous.
**Leaves** have blunt apexes

**Blue Douglas foliage** (below) set more radially.
**Leaves** blue-gray, thick, stand above the shoot in first year, produce little scent.
**Apexes** are rounder

**Bracts** point forwards

**Habit** slender, grows ragged later.
**Foliage** in pendent masses

**Branches** whorled, upswept when young, later heavy, level

**Blue Douglas cone** (right) smaller, 5 cm.
**Bracts** sometimes bent back.

**Douglas cone** (above) 8 cm, green when young

**Bracts** three-pronged, exserted, spreading or reflexed

**Bark** smooth, gray when young, later thick, ridged, corky, fissured

The genus *Pseudotsuga* was named after its resemblance to the Hemlocks (pp 62–3) but it also shows an affinity with *Abies* (pp 43–51). Douglas fir is native to the western side of the Rockies, but Blue douglas fir, recognizable by its cones and its blunt foliage, grows wild on the drier eastern side between Montana and Mexico. Large-coned Douglas fir (*P. macrocarpa*) is a Californian species with longer needles to 5 cm with bony acuminate tips and 10–15 cm cones with less exserted bracts.

61

# Hemlocks *Tsuga*

Hemlocks are a small group of conifers differing from spruces in their flattened needles, usually notched at apex, and the slender branchlets which lack the prominent *pulvini* of the *Picea*. Except in the Mountain hemlock, the cones are less than 3.5 cm.

## Western hemlock

### *Tsuga heterophylla*

**Needles** *2 cm at side but only 1 cm above, tip rounded*

**Cone** *3 cm, pendulous, ovoid, scales rounded, entire*

**Shoots** *slender, ribbed, cream-brown, hairs brown.*
**Buds** *small, ovoid, non-resinous*

**Leaf** *has white bands below*

**50 m. Crown** *conic, broader on old trees, dense, pendulous tips to straight branches ascending at 45°, low branches droop*

**Young tree** *(right) has leading shoot which arches over to form hanging curtain*

**Foliage** *spreading below, parted above*

**Bark** *thin, smooth, fissured and ridged in older trees.* **Bole** *straight, fluted; single stem straight from ground to tip of tree*

Western hemlock, a native of the western half of North America, is a fast-growing tree with attractive foliage, extremely tolerant of shade condition. The specific name refers to the irregular foliage arrangement, also a feature of *T. diversifolia*, a Japanese species, which has entire leaf margins, shorter leaves that are vividly white below and orange shoots.

# Canada hemlock

## *Tsuga canadensis*

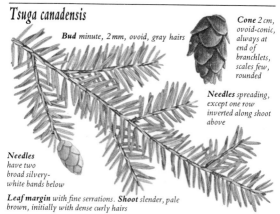

**Bud** *minute, 2 mm, ovoid, gray hairs*

**Cone** *2 cm, ovoid-conic, always at end of branchlets, scales few, rounded*

**Needles** *spreading, except one row inverted along shoot above*

**Needles** *have two broad silvery-white bands below*

**Leaf margin** *with fine serrations.* **Shoot** *slender, pale brown, initially with dense curly hairs*

Unlike *heterophylla* Canada hemlock usually has a multiple stem and grows in the eastern half of North America. Carolina hemlock (*T. carolinensis*), from the south-eastern USA, has entire margined needles on shiny orange shoots, and larger long-ovoid cones to 3.5 cm with thin rounded scales.

# Mountain hemlock

## *Tsuga mertensiana*

**Needles** *2.5 cm, slender, stomata on both sides*

**Leaves** *dense, green or gray*

**Leaves** *set all around shoot.* **Shoot** *hairy, shiny, pale brown.* **Bark** *deeply furrowed with rounded ridges, rough on younger trees and more orangey*

**30 m. Crown** *spire-like, tip nodding*

**Cone** *8 cm, cylindrical, tapers at both ends, scales rounded, reflexed when cone open*

Mountain hemlock is midway between a *Tsuga* and a *Picea*, but is distinguishable by the petiole-like base of its needle. It is used for timber, and has a similar though higher distribution to Western hemlock, with which it has hybridized to produce Jeffrey hemlock (*T. x jeffreyi*) with spreading partly radial needles.

# Pines *Pinus*

The members of this genus can be identified by their leaf groupings and divide into hard pines (Diploxylin) such as Scotch pine (p 64) and soft pines (Haploxylin) like Blue pine (p 76). The former have leaves in fascicles of two or three, rough bark and cones whose woody scales have central umbos; the latter have leaves in fives and softer cones with umbos at their scale tips.

## Scotch pine

③ 🌲 🌱 🌿 🌲

### *Pinus sylvestris*

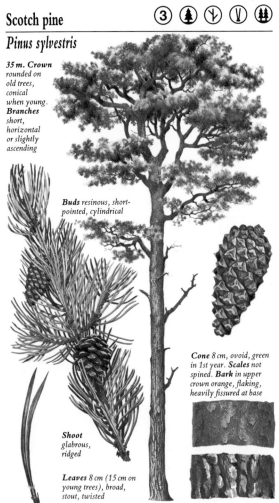

**35 m. Crown** rounded on old trees, conical when young. **Branches** short, horizontal or slightly ascending

**Buds** resinous, short-pointed, cylindrical

**Cone** 8 cm, ovoid, green in 1st year. **Scales** not spined. **Bark** in upper crown orange, flaking, heavily fissured at base

**Shoot** glabrous, ridged

**Leaves** 8 cm (15 cm on young trees), broad, stout, twisted

This tree has a wide natural range across Europe and Asia from the Atlantic to Pacific; in America it has naturalized locally in other areas from upstate New York across to the west coast. Its change in bark color and texture is distinctive, as is its gray-green to bright blue-green foliage; that of 'Aurea' is golden in winter. It is the main species cultivated for Christmas trees.

**Foliage** dense, in forward-pointing bunches. **Shoot** ridged and shiny

**Leaves** 15 cm, stiff, curved in 2nd year. **Sheaths** 1 cm, persistent

**Buds** long, 12 mm

**Bark** coarse, deeply furrowed. **Corsican pine** (below): 45 m. **Crown** lighter

**Cones** (both trees) 8 cm

**Corsican foliage** slender, spreading, twisted

**Leaves** 18 cm, in spaced fascicles. **Bud** resinous

Widely distributed throughout the Mediterranean, Black pine (*P. nigra*) occurs in several forms, two of which are shown here: Austrian pine is a hardy, densely crowned tree, usually growing on several stems to 30 m, Corsican pine is more vigorous and often planted for its timber. Their cones are identical but they can usually be identified by their shoots, buds and leaves. Bosnian pine (*P. leucodermis*) has similarly dense foliage but cobalt blue unripe cones and bloomed shoots.

# Shore pine • Lodgepole pine ④ 🌲 🌱 🌿 🌲🌲

## *Pinus contorta var. contorta • P. contorta var. latifolia*

**Shore pine leaves** *5 cm, straight.*
**Lodgepole leaves** *to 10 cm*

**Shore pine foliage** *dense*

**Lodgepole crown** *more open.* **Bark** *finely scaled*

**Cones** *(both trees) 5 cm, prickled*

**Lodgepole foliage** *less dense, splayed, twisted*

♂ **flowers** *at base of new shoots shedding pollen in April*

**30 m**

Shore pine is found along the Pacific coast while Lodgepole pine's range extends inland, hybridizing with Jack pine in some areas. Both these pines are short-lived and regenerate after fires.

# Stone pine ⑨ 🌲 🌱 🌿 🏛

## *Pinus pinea*

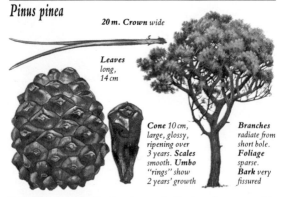

**20 m. Crown** *wide*

**Leaves** *long, 14 cm*

**Cone** *10 cm, large, glossy, ripening over 3 years.* **Scales** *smooth.* **Umbo** *"rings" show 2 years' growth*

**Branches** *radiate from short bole.* **Foliage** *sparse.* **Bark** *very fissured*

Instantly recognizable by its umbrella-shaped crown, this Mediterranean species has large, 2 cm, wingless seeds which have been a culinary delicacy since the time of the Romans.

# Jack pine

## *Pinus banksiana*

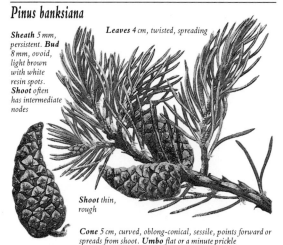

**Sheath** 5 mm, persistent. **Bud** 8 mm, ovoid, light brown with white resin spots. **Shoot** often has intermediate nodes

**Leaves** 4 cm, twisted, spreading

**Shoot** thin, rough

**Cone** 5 cm, curved, oblong-conical, sessile, points forward or spreads from shoot. **Umbo** flat or a minute prickle

Jack pine, an often ragged tree to 25 m, is found across Canada east of the Rockies and is unique in its forward-pointing cones. Table mountain pine (*P. pungens*), from the Appalachians, has similar foliage to 9 cm and cones with sharp hooked umbos.

# Virginia pine

## *Pinus virginiana*

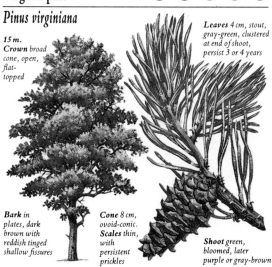

**15 m.** **Crown** broad cone, open, flat-topped

**Leaves** 4 cm, stout, gray-green, clustered at end of shoot, persist 3 or 4 years

**Bark** in plates, dark brown with reddish tinged shallow fissures

**Cone** 8 cm, ovoid-conic. **Scales** thin, with persistent prickles

**Shoot** green, bloomed, later purple or gray-brown

Virginia or Scrub pine is distinguishable from Jack pine by its cones, and is found in the Appalachians and the Atlantic states. Spruce pine (*P. glabra*), from southeastern USA, has slender, flexible leaves, globose cones to 5 cm and a gray furrowed bark.

## *Pinus muricata*

**Leaves** *15 cm, stiff, spreading, crowded, yellowish near base*

**Leaves** *green-gray*

**Cones** *8 cm, ovoid, oblique, sessile, often persist unopened clustered on shoot.*
**Scales** *on outer basal side larger, protrude, end in stout prickle*

**Sheath** *2 cm, persists.* **Bud** *1.5 cm, red-brown, cylindric, acute*

**Umbo** *spined*

**Shoot** *rough, glabrous*

Bishop pine forms a tree to 30 m with a domed crown and usually heavy branching, often profusely covered with closed cones. In its seven coastal and island sites in California it shows some variation, including a northern form which has a narrowly conic crown and darker, bluish-gray leaves.

# Shortleaf pine ⑥ 🌲 🌿 🌿 🌲 🌲

## *Pinus echinata*

**30 m. Crown** *narrow conic, later spreading*

**Leaves** *12 cm, in both 2s and 3s, slender, flexible.*
**Bud** *1 cm, gray-brown, slightly resinous*

**Cone** *6 cm, ovoid to oblong-conic.*
**Scales** *thin*

**Cone** *sessile or short-stalked.*
**Umbos** *sharp, often deciduous*

**Bark** *red-brown, scaly with resin pockets*

Shortleaf pine is a major lumber tree in the southeastern USA and may have sprouts on its branches. Slash pine (*P. elliottii*), from Florida, also has 2- and 3-needled fascicles, with glossy leaves to 30 cm, lustrous dark brown cones to 15 cm and papery scaly bark.

# Red pine

## *Pinus resinosa*

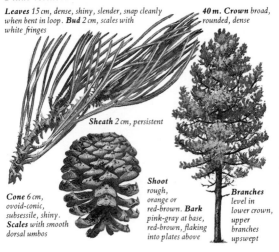

**Leaves** 15 cm, dense, shiny, slender, snap cleanly when bent in loop. **Bud** 2 cm, scales with white fringes

**40 m. Crown** broad, rounded, dense

**Sheath** 2 cm, persistent

**Cone** 6 cm, ovoid-conic, subsessile, shiny. **Scales** with smooth dorsal umbos

**Shoot** rough, orange or red-brown. **Bark** pink-gray at base, red-brown, flaking into plates above

**Branches** level in lower crown, upper branches upswept

Red pine is also misleadingly called Norway pine, though it is native from New England to Minnesota. It resembles Scotch pine in its flaking red-brown bark in the upper crown, but its whorled, brittle foliage and its lemon-scented resin are distinctive.

# Aleppo pine

## *Pinus halepensis*

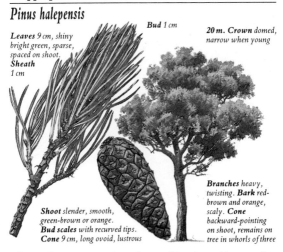

**Leaves** 9 cm, shiny bright green, sparse, spaced on shoot. **Sheath** 1 cm

**Bud** 1 cm

**20 m. Crown** domed, narrow when young

**Shoot** slender, smooth, green-brown or orange. **Bud scales** with recurved tips. **Cone** 9 cm, long ovoid, lustrous

**Branches** heavy, twisting. **Bark** red-brown and orange, scaly. **Cone** backward-pointing on shoot, remains on tree in whorls of three

This Mediterranean species, tolerant of dry sites with a very low summer rainfall, is frequently planted in California. Mondell pine (*P. eldarica*), native to Iran, Pakistan and Afghanistan, has forward-pointing cones on stout shoots with stiffer dark leaves.

# Maritime pine

## *Pinus pinaster*

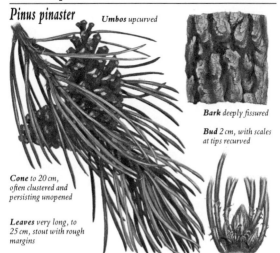

**Umbos** *upcurved*

**Bark** *deeply fissured*

**Bud** *2 cm, with scales at tips recurved*

**Cone** *to 20 cm, often clustered and persisting unopened*

**Leaves** *very long, to 25 cm, stout with rough margins*

This tree thrives on poor, sandy sites and is widely planted in Mediterranean countries for its resin. This is tapped by longitudinal wounds made in the bark and is used in turpentine manufacture. Its leaves are the longest and stoutest of any European pine.

# Pinyon

## *Pinus edulis*

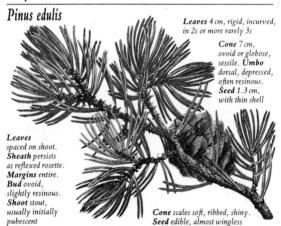

**Leaves** *4 cm, rigid, incurved, in 2s or more rarely 3s*

**Cone** *7 cm, ovoid or globose, sessile.* **Umbo** *dorsal, depressed, often resinous.* **Seed** *1.3 cm, with thin shell*

**Leaves** *spaced on shoot.* **Sheath** *persists as reflexed rosette.* **Margins** *entire.* **Bud** *ovoid, slightly resinous.* **Shoot** *stout, usually initially pubescent*

**Cone** *scales soft, ribbed, shiny.* **Seed** *edible, almost wingless*

The pinyons are a group of soft pines from the semi-arid regions of southwestern USA with several characteristics of the hard pines, such as dorsal umbos and nearly persistent sheaths. They form small, bushy trees to 10 m with rounded crowns and their edible seeds, known as pine nuts, are harvested commercially. Single-leaf pinyon (*P. monophylla*) bears single round gray-green leaves to 6 cm with white lines of stomata.

# Pitch pine

## *Pinus rigida*

**Bud** 2 cm, cylindric

**Leaves** 13 cm, rigid, stout, twisted, often at right angle to shoot

**20 m. Crown** open, irregular

**Cone** 8 cm, subsessile, ovoid-conic

**Shoot** stout, rough, gray-brown

**Bark** red-brown, coarsely fissured into plates

**Umbo** a curved prickle

**Bole** bears epicormic shoots

Pitch pine, from the northeastern USA, is unusual among pines in usually sprouting epicormic shoots from its bole. Pond pine (*P. serotina*), from southeastern USA, differs in having flexible needles to 20 cm, globose cones and thickly resinous buds.

# Loblolly pine

## *Pinus taeda*

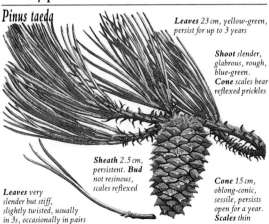

**Leaves** 23 cm, yellow-green, persist for up to 3 years

**Shoot** slender, glabrous, rough, blue-green.
**Cone** scales bear reflexed prickles

**Sheath** 2.5 cm, persistent. **Bud** not resinous, scales reflexed

**Leaves** very slender but stiff, slightly twisted, usually in 3s, occasionally in pairs

**Cone** 15 cm, oblong-conic, sessile, persists open for a year.
**Scales** thin

Loblolly pine is a densely-crowned tree to 35 m in the southeastern USA, where it covers several million acres. It is better suited to wet sites than most pines, and its name is derived from 'loblollies'—moist depressions. It is one of a group of southern pines raised for lumber, plywood and pulp.

# Longleaf pine

## *Pinus palustris*

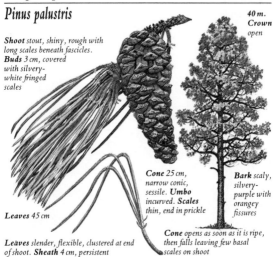

**40 m. Crown** open

**Shoot** stout, shiny, rough with long scales beneath fascicles.
**Buds** 3 cm, covered with silvery-white fringed scales

**Cone** 25 cm, narrow conic, sessile. **Umbo** incurved. **Scales** thin, end in prickle

**Bark** scaly, silvery-purple with orangey fissures

**Leaves** 45 cm

**Leaves** slender, flexible, clustered at end of shoot. **Sheath** 4 cm, persistent

**Cone** opens as soon as it is ripe, then falls leaving few basal scales on shoot

This species, native to southeastern USA, has the longest needles of any conifer, and can be recognized by its conspicuous buds, which appear to be covered with cobwebs. It thrives on dry sandy sites and is grown for its timber and resin.

# Monterey pine

## *Pinus radiata*

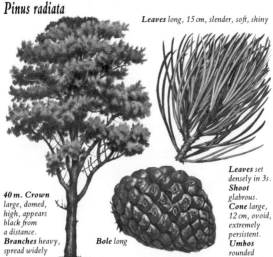

**Leaves** long, 15 cm, slender, soft, shiny

**Leaves** set densely in 3s. **Shoot** glabrous. **Cone** large, 12 cm, ovoid, extremely persistent. **Umbos** rounded

**40 m. Crown** large, domed, high, appears black from a distance.
**Branches** heavy, spread widely

**Bole** long

Native to California, Monterey pine's oblique cones may persist for over twenty years. Knobcone pine (*P. attenuata*) has assymetrical ovoid-conic cones, thin bark, smooth shoots and yellow-green leaves.

# Ponderosa pine
## *Pinus ponderosa*

**40 m. Crown** *variable*

**Leaves** *to 22 cm*

**Crown** *fairly open.* **Bole** *long, straight*

**Leaves** *dense, stout, usually in fascicles of 3.* **Sheath** *2 cm*

**Cone** *12 cm, when falls base left behind.* **Scales** *ridged, prickled*

This tree grows at varying altitudes and on dry sites throughout western North America. Its foliage is very variable—even on the same tree, leaves may be set in fascicles of both two and three.

# Jeffrey pine
## *Pinus jeffreyi*

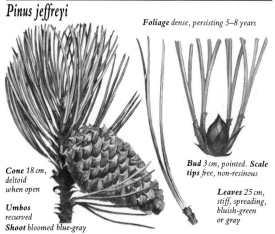

**Foliage** *dense, persisting 5–8 years*

**Cone** *18 cm, deltoid when open*

**Umbos** *recurved* **Shoot** *bloomed blue-gray*

**Bud** *3 cm, pointed.* **Scale tips** *free, non-resinous*

**Leaves** *25 cm, stiff, spreading, bluish-green or gray*

Native to Oregon and California and rarely exceeding 35 m, Jeffrey pine is smaller than Ponderosa but can grow at greater heights. Its distinguishing features are its bloomed shoot, and when available, its larger, broad-based cones.

# Coulter pine

## *Pinus coulteri*

**Cone** *huge, 35 cm, weighs up to 2.5 kg.*
**Seeds** *12 mm.* **Scales** *thick.*
**Umbos** *sharp*

**Foliage** *stout, rigid, spaced, not densely set, becomes crinkled in 2nd year*

**Shoot** *stout, glabrous*

**Leaves** *30 cm.* **Sheath** *2.5 cm*

While closely related to Ponderosa and Jeffrey pines, Coulter pine can be recognized by the size of its cones, normally set at the summit, or its longer hanging leaves. Digger pine (*P. sabiniana*) holds its 25 cm leaves level.

# Lacebark pine

## *Pinus bungeana*

**Foliage** *spaced, not set densely on shoot*

**Bark** *smooth, flaking through white, yellow, olive and purple to gray-green*

**Leaves** *8 cm, slender, smooth, finely toothed, in close fascicles.* **Sheath** *deciduous*

**Cone** *6 cm, on 2 cm stalk.* **Umbos** *dorsal, spined*

This species has a low, usually bushy crown and is cultivated for the splendor of its bark. This is initially smooth and gray-green and then flakes away to leave rounded white patches which turn through yellow, green, red and purple to purple-brown. In its native China, the bark of old trees turns a further chalk white.

# Swiss stone pine

## *Pinus cembra*

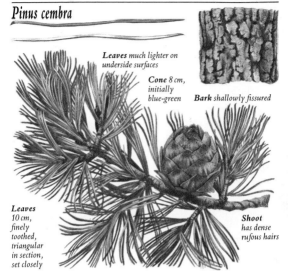

**Leaves** much lighter on underside surfaces

**Cone** 8 cm, initially blue-green

**Bark** shallowly fissured

**Leaves** 10 cm, finely toothed, triangular in section, set closely

**Shoot** has dense rufous hairs

Swiss stone or Arolla pine grows wild at high altitudes in the mountains of central Europe. The crown, to 25 m, is broad and dense. Korean pine (*P. koraiensis*) has longer 12 cm needles, a more open crown and 15 cm cones with thick, fleshy scales.

# Macedonian pine

## *Pinus peuce*

30 m

**Crown** dense

**Leaves** 12 cm, dense

**Leaves** dense, rigid

**Umbos** incurved

**Branches** whorled, very upswept in upper crown

**Cone** 15 cm, stalked. **Scales** thin, convex

Rarely planted for timber outside its native Albania, Bulgaria and Yugoslavia, this attractive tree is similar to Blue pine but has finer shoots, more curved cones and leaves which are shorter, denser, rigid and more forward-pointing.

# Blue pine

## *Pinus wallachiana*

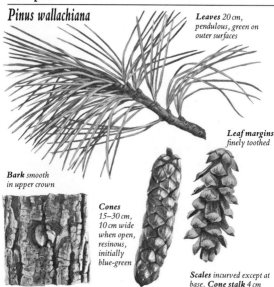

**Leaves** 20 cm, pendulous, green on outer surfaces

**Leaf margins** finely toothed

**Bark** smooth in upper crown

**Cones** 15–30 cm, 10 cm wide when open, resinous, initially blue-green

**Scales** incurved except at base. **Cone stalk** 4 cm

Blue pine, often known as Bhutan pine, is indigenous throughout the Himalayas. It has been widely planted as an ornamental and forms a broad, heavily branched tree to 35 m, with stout, glabrous shoots and curved leaves. Mexican white pine (*P. ayacahuite*) has straight needles and more tapered cones.

# Armand pine

## *Pinus armandii*

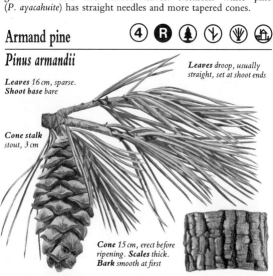

**Leaves** 16 cm, sparse. **Shoot base** bare

**Leaves** droop, usually straight, set at shoot ends

**Cone stalk** stout, 3 cm

**Cone** 15 cm, erect before ripening. **Scales** thick. **Bark** smooth at first

Dedicated to its discoverer, Père Armand David, this pine has a very wide distribution across China and in Burma. It resembles Blue pine but has finer shoots and barrel-shaped cones.

# Limber pine

## *Pinus flexilis*

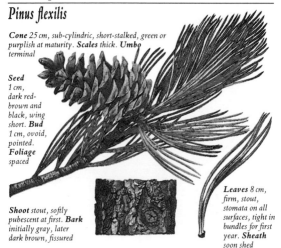

**Cone** *25 cm, sub-cylindric, short-stalked, green or purplish at maturity.* **Scales** *thick.* **Umbo** *terminal*

**Seed** *1 cm, dark red-brown and black, wing short.* **Bud** *1 cm, ovoid, pointed.* **Foliage** *spaced*

**Shoot** *stout, softly pubescent at first.* **Bark** *initially gray, later dark brown, fissured*

**Leaves** *8 cm, firm, stout, stomata on all surfaces, tight in bundles for first year.* **Sheath** *soon shed*

A broad-crowned tree to 15 m, Limber pine is native at heights of up to 3,500 m on both sides of the Rocky Mountains from Canada to Mexico. The similar Whitebark pine (*P. albicaulis*) has ovoid cones to 7 cm which do not open but disintegrate and fall.

# Sugar pine

## *Pinus lambertiana*

60 m

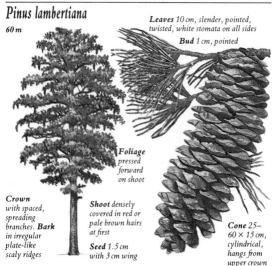

**Leaves** *10 cm, slender, pointed, twisted, white stomata on all sides*

**Bud** *1 cm, pointed*

**Foliage** *pressed forward on shoot*

**Crown** *with spaced, spreading branches.* **Bark** *in irregular plate-like scaly ridges*

**Shoot** *densely covered in red or pale brown hairs at first*

**Seed** *1.5 cm with 3 cm wing*

**Cone** *25–60 × 15 cm, cylindrical, hangs from upper crown*

Sugar pine, named after the sugary sap or resin which exudes from wounds, is the tallest pine and produces the largest cones, although they are lighter than those of Coulter pine (p74). It is restricted to Oregon and California, and is prone to Blister rust.

# Japanese white pine

## *Pinus parvifolia*

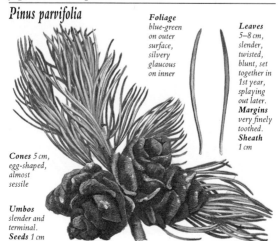

**Foliage**
blue-green
on outer
surface,
silvery
glaucous
on inner

**Leaves**
5–8 cm,
slender,
twisted,
blunt, set
together in
1st year,
splaying
out later.
**Margins**
very finely
toothed.
**Sheath**
1 cm

**Cones** 5 cm,
egg-shaped,
almost
sessile

**Umbos**
slender and
terminal.
**Seeds** 1 cm

In its wild form, this species grows to a height of 25 m but is more usually encountered as a lower, slow-growing tree that rarely reaches 10 m. It has a wide crown and tiered branches and was probably developed for Japanese ornamental gardens. The leaves of both types are the most twisting of any pine.

# Western white pine

## *Pinus monticola*

**Cones** 35 cm, tapering, slightly curved,
set on short stalks, numerous.
**Scales** thin, rounded, resinous

**Foliage** dense,
set in tufts

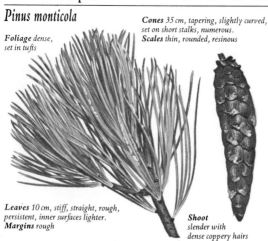

**Leaves** 10 cm, stiff, straight, rough,
persistent, inner surfaces lighter.
**Margins** rough

**Shoot**
slender with
dense coppery hairs

Native to the Pacific coast and in mountains as far inland as Montana, *P. monticola* has a dense crown and can reach heights of 50 m. Like all soft pines, the bark of young trees and that high up on old trees is smooth and gray-green and often attacked by blister rust, a fungal disease which affects all American white pines and kills many old trees.

# Eastern white pine

## *Pinus strobus*

**Leaves** *12 cm, paler on inner surfaces*

**Shoot** *ridged behind leaves, only hairy on ridges*

**Cones** *15 cm, narrow, pointed*

**Scales** *convex*

**30 m. Crown** *conic when young, later flat-topped.* **Bark** *narrowly fissured*

Native to the forests of the eastern USA, Eastern white pine forms a tree to 40 m, often on several stems, and differs from Western white pine in its glabrous shoots and shorter, pointed cones which are only 15 cm high. The leaves are stiffer and only persist for two seasons. It is susceptible to white pine blister rust.

# Bristlecone pine

## *Pinus aristata*

**Cone** *8 cm, spreading, with short stalk.*
**Umbo** *has 6 mm spreading or curving prickle*

**Bud** *7 mm, pointed, scales free at base*

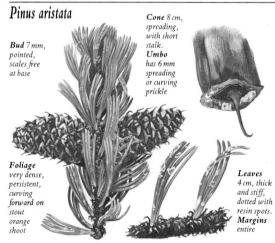

**Foliage** *very dense, persistent, curving forward on stout orange shoot*

**Leaves** *4 cm, thick and stiff, dotted with resin spots.* **Margins** *entire*

This is the commonest of three trees from the southwestern USA known as "Foxtail pines" because of their persistent, dense foliage. Foxtail pine (*P. balfouriana*) has longer, unspotted leaves, tight buds and 13 cm cones with minute umbos, while Ancient pine (*P. longaeva*) has foxtail foliage and Bristlecone cones.

# The broadleaf trees

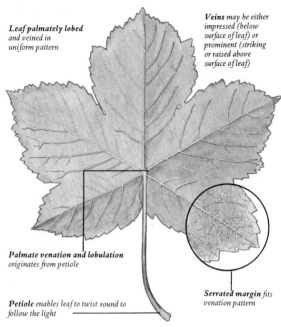

**Leaf palmately lobed** and veined in uniform pattern

**Veins** may be either impressed (below surface of leaf) or prominent (striking or raised above surface of leaf)

**Palmate venation and lobulation** originates from petiole

**Petiole** enables leaf to twist round to follow the light

**Serrated margin** fits venation pattern

The trees known as the broadleaf trees, because their leaves have broad blades, are more properly classified as the Angiospermae, so called because their seeds develop from ovules which are enclosed in an ovary. Many can be classified as Dicotyledonae, because their seedlings have two seed leaves, but some are Monocotylonedae, with one seed leaf, which go on to produce leaves with long unbranched veins and less complexly branched crowns than the Dicotyledonae.

The shape of the leaf, the best feature for identification purposes, is governed by its composition and function, a vital element being the venation system, illustrated above in a Plane-tree maple leaf. The veins strengthen the thin blade, and the many complex patterns they form are useful for identification. Their main function, however, and that of the leaf, is to produce energy for the tree by photosynthesis—the fixing (i.e., turning into solid form) of carbon from the air. This process has to take place in daylight and needs water, which is transported up from the root system to spread across the surface of the leaf blade through the network of veins. The veins are two-way channels; they transport water and nutrients to the leaf, and carry the sugar sap, the end product of photosynthesis, to the rest of the tree.

After assisting in photosynthesis, the water from the roots transpires or evaporates into the atmosphere, something which happens to a far lesser extent in most conifers because the thick waxy surface of their thin needles retains moisture. Temperate broadleaf trees are nearly all deciduous, dropping all their leaves each winter. Once the temperature drops, the tree cannot extract from the cold soil enough water to maintain transpiration, and freezing temperatures would damage the leaves. The nutrient material is withdrawn from the leaves before they wither and die.

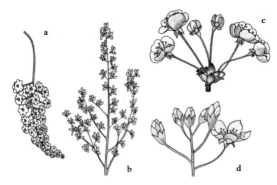

## Flowers

The broadleaf trees flower for a relatively short time, but these flowers are very useful in identification. While conifers transfer the pollen from the male to the female flowers on the wind, most broadleaves are insect pollinated, and their flowers have brightly colored petals and strong scents to attract the insects. Four types of flower arrangement are illustrated above. A **raceme (a)** is a simple group of stalked flowers on a long single rachis. A **panicle (b)** is a looser compound or branched flower cluster. An **umbel (c)** is an inflorescence with pedicels all rising from the same point. A **corymb (d)** is a flat-topped flower cluster in which the outside flowers open first.

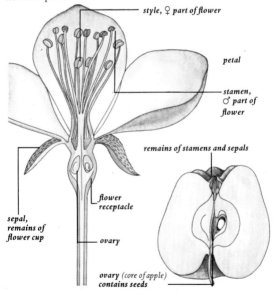

*style, ♀ part of flower*

*petal*

*stamen, ♂ part of flower*

*remains of stamens and sepals*

*flower receptacle*

*sepal, remains of flower cup*

*ovary*

*ovary (core of apple) contains seeds*

One mode of the transition from flower to fruit is illustrated with the apple above. The flesh develops from the flower receptacle, and once the petals have dropped, the stamens and sepals wither until their remains are at the base of the fruit. What was the ovary becomes the core of the apple, which contains the seeds.

81

## Fruit

The basic function of the fruit is to protect the seeds as they develop, until they are ready for dispersal to propagate the species. The fruit may be sparser than the flowers, either because some of the flowers did not develop or because the majority of the inflorescence was male. There are several different types of fruit: the apple illustrated on the previous page is a **pome**, which has a fleshy covering around one or more seeds in several fused cells; and some other examples are illustrated below. A **capsule (a)**, as borne by eucalyptus trees, is a hard woody pod which releases many small seeds through 3–6 terminal seeds. A **drupe (b)**, such as a cherry for example, has a fleshy exterior around a stone which contains one or more seeds. A **cone-like structure (c)**, as borne by the alders, has spirally or oppositely arranged scales which each carry two or more seeds. A **samara (d)**, as borne by ashes, has a single seed in a case on the end of a long wing. A **nut (e)**, such as the acorn, bears its seeds inside a hard shell; the acorn is set in a cup called an **involucre** or **cupule**.

## Seed germination

The few seeds which alight on suitable ground can lie dormant for one or two years before germination. They are gradually softened by moisture from the soil, and eventually a tiny root is produced by cell division, which breaks through the seed coat and penetrates the ground. It next develops tiny hairs to extract water and nutrients from the earth; these are renewed every year throughout its life. Sometimes two seed leaves withdraw from the withered seed casing as the seedling, the first plant, is formed. In the Horse chestnut, illustrated below, they remain inside.

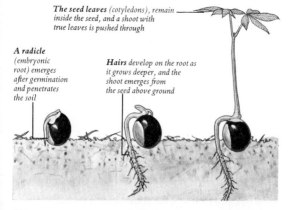

*The seed leaves* (cotyledons), remain inside the seed, and a shoot with true leaves is pushed through

*A radicle* (embryonic root) emerges after germination and penetrates the soil

*Hairs* develop on the root as it grows deeper, and the shoot emerges from the seed above ground

# Willow family Salicaceae

This family has 350 or so species of willow and poplar which are mainly natives of the northern hemisphere. The chief feature uniting them all is their flowers. These have neither petals nor sepals and are borne in catkins which usually appear with or before their tree's new leaves and are either male or female. Only one type is usually carried. Both willows and poplars prefer moist sites and hybridize so easily that positive identification is not always easy.

## Poplars *Populus*

Poplars have wind-pollinated catkins and leaves whose broad blades have long petioles which may be flattened at one end. Their deciduous shoots bear terminal buds and all their buds have overlapping scales. Trees which have hybridized grow more vigorously and provide better timber than their parents.

# White poplar

## *Populus alba*

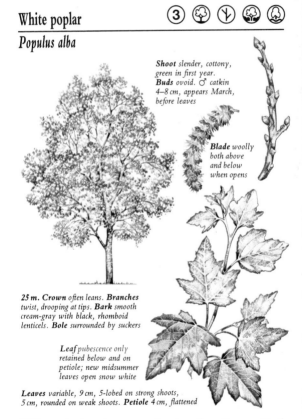

**Shoot** *slender, cottony, green in first year.* **Buds** *ovoid.* ♂ *catkin 4–8 cm, appears March, before leaves*

**Blade** *woolly both above and below when opens*

**25 m. Crown** *often leans.* **Branches** *twist, drooping at tips.* **Bark** *smooth cream-gray with black, rhomboid lenticels.* **Bole** *surrounded by suckers*

**Leaf** *pubescence only retained below and on petiole; new midsummer leaves open snow white*

**Leaves** *variable, 9 cm, 5-lobed on strong shoots, 5 cm, rounded on weak shoots.* **Petiole** *4 cm, flattened*

White poplar is native to central and southern Europe, North Africa and central Asia and, with Quaking aspen (p 85), belongs to a group of poplars with smooth barks and lobed or coarsely serrate leaves. 'Pyramidis' is a fastigiate clone, broader than Lombardy poplar (p 86); 'Richardii' has golden yellow leaves.

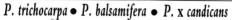

## *P. trichocarpa* ● *P. balsamifera* ● *P. x candicans*

*Western balsam: ♂ catkin (right) long and thick, 8 × 1.5 cm, appearing in April before leaves.* **Buds** *2 cm, adpressed, resinous, slightly pubescent.* **Fruit** *(above) 3-valved, hairy, cottony by May when it is shed*

*Western balsam (P. trichocarpa):* 35 m. **Crown** conic. **Branches** whorled, numerous. **Leaves** (right) large, 15–30 cm, thick, yellow in autumn, always yellowish-white below. **Petioles** 3 cm, stout round

*Eastern balsam (P. balsamifera):* **Shoot** (right) round, not angled. **Buds** longer than those of Western balsam, adpressed. **Leaves** (below) 12 cm, broader, more obtuse than Western balsam but very variable. **Blades** glabrous, slightly downy below, finely serrate. **Petioles** 7 cm, round

*Balm of Gilead poplar* (*P. x candicans*): **Leaves** *(above)* 15 cm, downy beneath. **Margins** ciliate. **Petioles** 7 cm, downy, roundish. **Shoot** hairy initially, angular. *'Aurora' summer leaves (top) develop cream-white and pink marbling*

These trees are recognizable by the balsamiferous odor pervaded by their large, resinous buds and new foliage in early summer. Their leaves are always yellow-white below and have neither flattened leaf stalks nor translucent margins. Western and Eastern balsams are both native to North America and are fast-growing, the latter putting out suckers around its bole. Balm of Gilead poplar produces these so easily and profusely that they soon become a nuisance, and since its trunk is also prone to bacterial canker, its planting is not recommended; the 'Aurora' form, with variegated foliage, is preferable. The Chinese balsam poplar (*P. szechuanica*) is notable for its very large leaves whose undersides are initially downy and purple.

# Quaking aspen

## Populus tremuloides

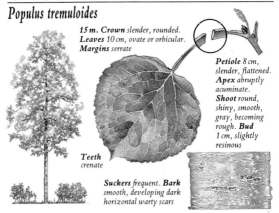

**15 m. Crown** slender, rounded.
**Leaves** 10 cm, ovate or orbicular.
**Margins** serrate

**Petiole** 8 cm, slender, flattened.
**Apex** abruptly acuminate.
**Shoot** round, shiny, smooth, gray, becoming rough. **Bud** 1 cm, slightly resinous

**Teeth** crenate

**Suckers** frequent. **Bark** smooth, developing dark horizontal warty scars

Named after the way its leaves flutter on their long petioles in even the slightest breeze, Quaking aspen is found across the north and down the Rockies. Flowers and fruit are borne in pendent 10 cm catkins. Big-tooth aspen (*P. grandidenta*), from the northeast, has coarsely toothed 15 cm leaves and hairy buds.

# Eastern cottonwood

## Populus deltoides

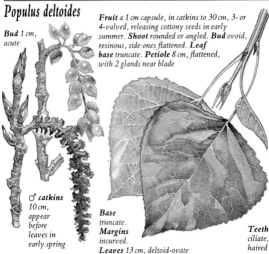

**Bud** 1 cm, acute

**Fruit** a 1 cm capsule, in catkins to 30 cm, 3- or 4-valved, releasing cottony seeds in early summer. **Shoot** rounded or angled. **Bud** ovoid, resinous, side ones flattened. **Leaf base** truncate. **Petiole** 8 cm, flattened, with 2 glands near blade

♂ **catkins** 10 cm, appear before leaves in early spring

**Base** truncate.
**Margins** incurved.
**Leaves** 13 cm, deltoid-ovate

**Teeth** ciliate, haired

Growing to 40 m, with a spreading crown and gray-brown, deeply furrowed bark, Eastern cottonwood is found throughout the eastern USA, and with *P. nigra* (p86) is a parent of most of the hybrid Black poplars. Plains cottonwood (*P. sargentii*) has slightly hairy buds and smaller, coarsely toothed leaves while Swamp cottonwood (*P. heterophylla*), from the east coast and the Mississippi basin, has shaggy bark and cordate ovate leaves.

## *Populus nigra*

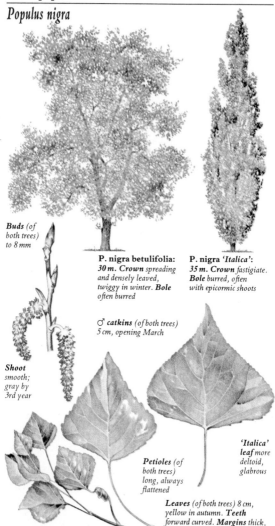

**Buds** (of both trees) to 8 mm

**P. nigra betulifolia:** *30 m.* **Crown** *spreading and densely leaved, twiggy in winter.* **Bole** *often burred*

**P. nigra 'Italica':** *35 m.* **Crown** *fastigiate.* **Bole** *burred, often with epicormic shoots*

♂ **catkins** (of both trees) *5 cm, opening March*

**Shoot** *smooth; gray by 3rd year*

'Italica' **leaf** *more deltoid, glabrous*

**Petioles** (of both trees) *long, always flattened*

**Leaves** (of both trees) *8 cm, yellow in autumn.* **Teeth** *forward curved.* **Margins** *thick, translucent*

The black poplar normally encountered in west Europe is *P. nigra* var. *betulifolia* and differs from its rarer continental type (*P. nigra*) in having birch-like leaves and initially downy twigs and leaf stalks. Its burred bole is diagnostic. The more common Lombardy poplar (*P. n.* 'Italica') may have arisen in central Asia rather than northern Italy as its name implies. The similar 'Plantierensis' clone has a leafier, slightly broader crown and initially pubescent petioles and shoots, nearly glabrous by midsummer. *Pemphigus bursarinus* aphids often attack these poplars and produce distinctively spiralled galls on their leaf stalks.

## *Populus* 'Androscoggin' ● *P.* 'Strathglass' ● *P.* 'Eugenii'

**'Androscoggin' leaves** (above) 13 cm, ovate or elliptic, dark green above, oily white below. **Apex** twisted, short acuminate. **Margins** finely serrate. **Teeth** crenate, ciliate. **Veins** shallowly impressed on upper surface. **Bud** 1.4 cm, resinous

**Shoot** round, slightly ridged near tip on vigorous shoots

**Lenticels** linear
**Shoot** glossy pale brown, slender, round

**Veins** impressed

**Apex** acuminate

**'Strathglass' leaves** (right) 6 cm, ovate to elliptic, smooth, lustrous above, whitish green below. **Petiole** 5 cm, slightly flattened, pale yellowish green. **Margins** slightly wavy, serrate. **Teeth** crenate, more coarsely so on young or vigorous shoots, often more so toward apex

**Bud** 1 cm, acute, lanceolate, slightly resinous. **Shoot** angled near tip on strong shoots

**Petiole** pinkish above, flattened toward blade

**'Eugenii' leaves** (left) 10 cm, broad ovate, shiny on both surfaces, getting thicker toward apex. **Shoot** slender, round, angled on vigorous shoots, marked with elliptic or round lenticels to 2 mm. **Bud** 5 mm, ovoid, acute, adpressed, slightly resinous

**Margins** serrate. **Teeth** gland-tipped, crenate, coarser toward apex

'Androscoggin', a fast-growing tree which soon attains 30 m and has an open, rather gaunt crown, is a hybrid between Western balsam (p84) and the Asiatic *P. maximowiczii*, which gives it the broad leaves with impressed veins. 'Strathglass' makes a denser tree with pendent lower branches—its parents are Black poplar (p86) and the Siberian *P. laurifolia*. 'Eugenii', a hybrid of Black poplar with Eastern cottonwood (p85), has a narrow crown with stiff, ascending branches.

# Willows *Salix*

Unlike poplars, willows are insect-pollinated, having stiffer, nectar-bearing catkins as well as longer and narrower stipulate leaves, shorter petioles and single-scaled buds. Their shoots lack terminal buds and growth occurs laterally behind the tips.

## White willow • Silver willow ③

### *Salix alba* • *S. alba* 'Sericea'

♂ catkin
(of both
willows)
5 cm

**Crown** rounded,
densely leaved

**White willow** (above):
25 m. **Branches** spread.
**Bole** stout. **Bark** cross-
ridged, fissured. **Leaves**
(right) 8 cm, open light
green. **Blades** have long,
white silky hairs above,
densely tomentose on
undersides

**Silver willow**
(above): 15 m. **Crown**
conic. **Branches**
ascend. **Branchlets** are
pendulous. **Leaves**
(below) similar to
White willow but
slightly wider and more
hairy. **Shoots and buds**
are also hairy and
appear silvery white

**Teeth** fine. **Leaf stalk**
short, 5 mm. **Shoot** fine,
densely haired, not brittle at
base. **Buds** 2 mm, flattened

Native to most of Europe, this vigorous species is one of the tallest willows and while sometimes confused with Crack willow (*S. fragilis*), can be distinguished by its less fragile twigs and less deeply fissured bark. It is often pollarded to produce pliant shoots which are used for basketry. The widely distributed 'Sericea' is readily identified by its coloring and more hairy foliage.

# Weeping willow

## *Salix x chrysocoma*

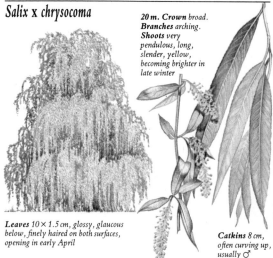

**20 m. Crown** broad. **Branches** arching. **Shoots** very pendulous, long, slender, yellow, becoming brighter in late winter

**Leaves** 10 × 1.5 cm, glossy, glaucous below, finely haired on both surfaces, opening in early April

**Catkins** 8 cm, often curving up, usually ♂

Attractive beside water toward which its foliage tends to bend, this is a hybrid of White willow and Chinese weeping willow (*S. babylonica*) which is far less common and has pendulous *brown* shoots and leaves with fewer teeth.

# Corkscrew willow

## *Salix matsudana* 'Tortuosa'

**Leaves** open in March

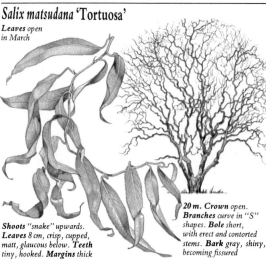

**20 m. Crown** open. **Branches** curve in "S" shapes. **Bole** short, with erect and contorted stems. **Bark** gray, shiny, becoming fissured

**Shoots** "snake" upwards. **Leaves** 8 cm, crisp, cupped, matt, glaucous below. **Teeth** tiny, hooked. **Margins** thick

Useful on dry soils, this willow makes an attractive garden tree. Its leaves are among the first to appear and the last to fall; in winter its distinctively contorted crown, which can only be confused with that of the Corkscrew hazel, is very striking.

# Scouler willow

## *Salix scouleriana*

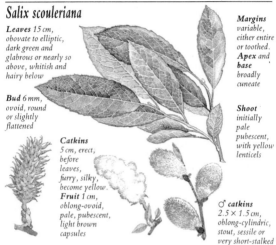

**Leaves** 15 cm, obovate to elliptic, dark green and glabrous or nearly so above, whitish and hairy below

**Bud** 6 mm, ovoid, round or slightly flattened

**Catkins** 5 cm, erect; before leaves, furry, silky, become yellow. **Fruit** 1 cm, oblong-ovoid, pale, pubescent, light brown capsules

**Margins** variable, either entire or toothed. **Apex** and **base** broadly cuneate

**Shoot** initially pale pubescent, with yellow lenticels

♂ **catkins** 2.5 × 1.5 cm, oblong-cylindric, stout, sessile or very short-stalked

This species is the tallest-growing of the 'Pussy willow' group, attaining 15 m at times, although it is as often a shrub, and is a native of the Rockies and the Pacific coast in differing forms. The male trees are very showy in flower.

# Black willow

## *Salix nigra*

**Catkins** terminal on current year's leafy shoots

**40 m,** but sometimes a shrub. **Crown** broad, on divided bole

**Shoot** red-brown, initially pubescent, soon glabrous, snaps easily at joints if bent

**Leaves** green both sides

**Bole** twisted

**Branches** very upright. **Bark** black or dark brown, rigid, deeply furrowed into scaly ridges. **Leaves** on short petiole

**Leaves** 15 × 2 cm, lanceolate, shiny above, dull below. **Apex** long-pointed. **Margins** serrate

This is the largest of the willows, and is found throughout the eastern USA. The fruit is a 1 cm green capsule. Pacific willow (*S. lasiandra*), from the west, has shorter leaves, gray-green below, and broad ovate 6 mm buds.

# Walnut family Juglandaceae

This is a group of large trees with pinnate leaves and pendent male catkins exposed over winter. The fruit, usually edible, is a drupe or nut and in *Juglans* and *Pterocarya* the pith is chambered.

## English walnut

### Juglans regia

**Leaves** 45 cm, pinnate.
**Leaflets** (5–9) to 18 cm.
**Catkins** before leaves

**Leaf margins** entire

**Leaflets** subsessile, except terminal

**Fruit** ripe by September, husk withering to expose hardened nut

**Fruit** 5 cm. **Seed** wrinkled

**25 m. Crown** wide, rounded. **Branches** low, very sinuous, heavy

**Bark** in smooth plates between wide fissures. **Bole** upright

**Shoot** stout. **Pith** chambered. **Buds** 6 mm, aromatic when rubbed

English, or Persian, Walnut has become naturalized in Europe and across Asia and been cultivated since the earliest times, although its exact origin is unknown. Large orchards of it exist in France and California. The fruit is either picked for pickling before the end of July while still fleshy or left until autumn when it has hardened. *J. regia* is the only walnut with entire leaflets.

# Black walnut

## *Juglans nigra*

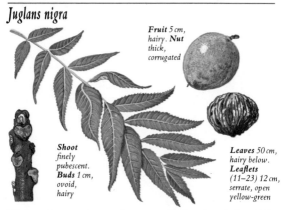

**Fruit** 5 cm, hairy. **Nut** thick, corrugated

**Shoot** finely pubescent. **Buds** 1 cm, ovoid, hairy

**Leaves** 50 cm, hairy below. **Leaflets** (11–23) 12 cm, serrate, open yellow-green

Black walnut, native to the eastern USA where it is widely planted for its timber and nuts, grows to 30 m with a broad open crown and dark brown or black cross-furrowed bark. It is distinguishable by its many leaflets, often without a terminal.

# Butternut

## *Juglans cinerea*

**Fruit** 6 cm, ovoid, pointed, husks with clammy matted hairs and 2 or 4 sutures. **Seed** 6 cm, oily, deeply ribbed

**Fruit** in racemes of 1–5

**Leaves** 75 cm. **Leaflets** 11–17, to 7.5 × 5 cm, lanceolate, finely serrate, pointed, sessile

**Rachis** stout, downy

**Leaflets** finely pubescent below. **Pith** chambered

**Shoot** stout, with reddish hairs at first; ridge of hairs remains above leaf. **Terminal bud** 2 cm, flattened

**30 m. Crown** broad, open dome. **Bole** short. **Bark** thick, light gray, furrowed

The sweet oily kernel which gives Butternut its name soon goes rancid, but is edible if picked early. This species grows in the eastern USA and into Canada, and is distinguished from Black walnut by its larger leaves and its clustered ovoid fruits.

# Hickories Carya

Hickories, restricted to North America and China, differ from walnuts in having solid pith, catkins in threes on a common peduncle and a fruit with a four-valved husk and a smooth seed.

## Pecan

### *Carya illinoensis*

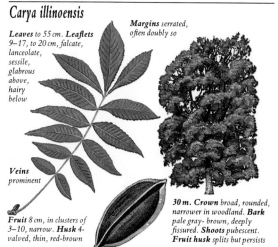

**Leaves** to 55 cm. **Leaflets** 9–17, to 20 cm, falcate, lanceolate, sessile, glabrous above, hairy below

**Margins** serrated, often doubly so

**Veins** prominent

**Fruit** 8 cm, in clusters of 3–10, narrow. **Husk** 4-valved, thin, red-brown

**30 m. Crown** broad, rounded, narrower in woodland. **Bark** pale gray-brown, deeply fissured. **Shoots** pubescent. **Fruit husk** splits but persists

Pecan is native to the Mississippi valley, but its range has been extended by widespread cultivation in the southern USA, where its nuts are an important commercial crop, and where pruning often alters the shape of the crown. It occasionally grows to 55 m.

## Mockernut hickory

### *Carya tomentosa*

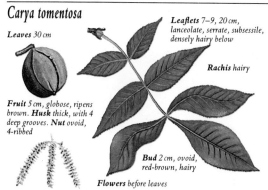

**Leaves** 30 cm

**Leaflets** 7–9, 20 cm, lanceolate, serrate, subsessile, densely hairy below

**Rachis** hairy

**Fruit** 5 cm, globose, ripens brown. **Husk** thick, with 4 deep grooves. **Nut** ovoid, 4-ribbed

**Bud** 2 cm, ovoid, red-brown, hairy

**Flowers** before leaves

♂ **catkins** slender, drooping in 3s on single stalk, pale yellow in summer

Mockernut, a native of the southeastern USA, forms a tree to 25 m with a broad open crown, distinguished by its fragrant foliage. Black hickory (*C. texana*) has black bark, 5–7 dark green leaflets and acuminate terminal buds never longer than 12 mm.

# Bitternut

## *Carya cordiformis*

**Leaflets** *serrate*

**25 m.** *Crown more open in older trees*

**Branches** *slender.* **Bark** *gray, fissured*

**Leaves** *25 cm.* **Fruit** *4 cm, encloses 4-ribbed nut*

**Shoot** *slender with white lenticels.* **Buds** *to 2.5 cm, slender, yellow, pubescent*

A species with a wide natural range throughout the eastern USA, this hickory grows to 20 m and is identifiable by the taste of its seeds and its yellow winter buds. Its leaves have up to 11 leaflets that are paler below and turn yellow in autumn. Pignut (*C. glabra*) has unribbed seeds and leaves with 5 glabrous leaflets.

# Shagbark hickory

## *Carya ovata*

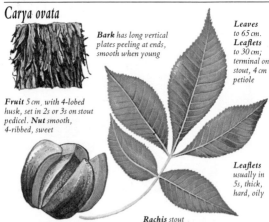

**Bark** *has long vertical plates peeling at ends, smooth when young*

**Leaves** *to 65 cm.* **Leaflets** *to 30 cm; terminal on stout, 4 cm petiole*

**Fruit** *5 cm, with 4-lobed husk, set in 2s or 3s on stout pedicel.* **Nut** *smooth, 4-ribbed, sweet*

**Leaflets** *usually in 5s, thick, hard, oily*

**Rachis** *stout*

This species is instantly identifiable by the nature of its bark which begins to flake into long, curling plates when the tree is about 25 years old. Shellbark hickory (*C. laciniosa*) has similar mature bark but its leaves have 7 leaflets with downy undersides.

# Birch family Betulaceae

A group of trees and shrubs comprising over 40 species, the members of this family carry male and female catkins on the same tree, the females being small, initially erect when flowering, and held above the males. In *Betula* (pp 95–7), *Alnus* (pp 98–9) and *Corylus* (pp 99–100), the males develop in autumn and are carried over winter, opening to release their pollen before, or when, the leaves appear in spring. Those of *Carpinus* (p 101), although pre-formed, are hidden in the winter buds.

## European white birch

### Betula pendula

**Leaves** *7 cm, doubly serrate*

**Leaves** *roughish, glabrous below*

*♂ catkin 3 cm, opens March–April.*
*♀ catkin (right) 3 cm, ripens brown*

**Bole** *fluted*

**30 m. Crown** *narrow.*
**Branches** *ascend when young, later very pendulous.* **Bark** *smooth, red-brown on young trees*

**Buds** *4 mm.*
**Shoot** *warty*

**'Dalecarlica':**
**Leaves** *5 cm.*
**Margins** *deeply dissected.*
**Petioles** *4 cm, slender*

**Immature** *♂ catkins (above) 2 cm, hang at shoot tip.* **Shoot** *glabrous, zig-zagged*

**Seed** *(above) 2 mm, ripe by Sept, has 2 large wings.* **Bract** *7 mm, 3-lobed*

**'Dalecarlica': Shoots** *long, very pendulous*

With a wide natural distribution throughout most of Europe and Asia Minor. European white birch is recognizable by its distinctive bark and hairless twigs. It thrives on light, dry and sandy soils, dislikes shade and is plentiful on heaths and moorlands. 'Dalecarlica', from southern Sweden, is one of its graceful cultivars.

# River birch

## *Betula nigra*

**Bud** 6 mm, covered with pale hairs in summer, glabrous, shiny brown in winter

**Fruit** 4 cm, erect, cylindric

**Shoot** densely hairy at first, becoming shiny red-brown by first winter, later dull and darker

**Fruit** a strobile on stout 1 cm hairy peduncle. **Bract** with narrow, erect, ciliate lobes

**Leaves** 8 cm, ovate, midrib hairy below. **Base** cuneate. **Petiole** hairy, flattened

**Leaves** lustrous above. **Apex** acute. **Margins** coarsely doubly serrate

**Bark** peels in papery flakes; black but pale pink-brown in fissures

This species is native to lowland sites in the southern and eastern USA, and is the only birch at all in the extreme southeast. It grows to 25 m with an irregular crown, usually dividing into two or three stems, and can tolerate prolonged flooding.

# Gray birch

## *Betula populifolia*

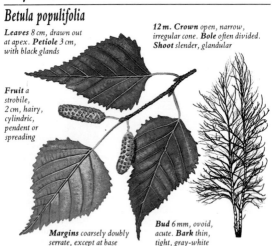

**Leaves** 8 cm, drawn out at apex. **Petiole** 3 cm, with black glands

**12 m. Crown** open, narrow, irregular cone. **Bole** often divided. **Shoot** slender, glandular

**Fruit** a strobile, 2 cm, hairy, cylindric, pendent or spreading

**Margins** coarsely doubly serrate, except at base

**Bud** 6 mm, ovoid, acute. **Bark** thin, tight, gray-white

Gray birch is very similar to Paper birch but can be identified by the long tail-like apex of its leaf and its tight, rather than peeling, bark. It is restricted to the northeast and springs up rapidly in barren soils, such as abandoned farmland.

# Paper birch

## *Betula papyrifera*

**Leaves** *10 cm but variable, doubly serrate, thick, very mat above.* **Petiole** *3 cm, stout, pubescent*

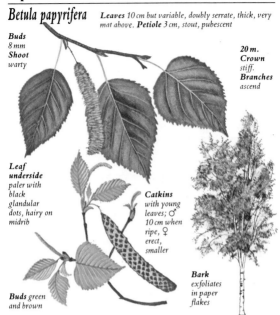

**Buds** *8 mm* **Shoot** *warty*

**20 m. Crown** *stiff.* **Branches** *ascend*

**Leaf underside** *paler with black glandular dots, hairy on midrib*

**Catkins** *with young leaves;* ♂ *10 cm when ripe,* ♀ *erect, smaller*

**Bark** *exfoliates in paper flakes*

**Buds** *green and brown*

Paper birch grows throughout Canada and northern USA; its creamy or pinkish white bark was once used by Indians to cover their canoes. Szechuan birch (*B. platyphylla* var. *szechuanica*) has a chalky white bark and leathery, 12 cm leaves, glaucous below.

# Yellow birch

## *Betula lutea*

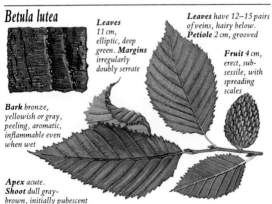

**Leaves** *11 cm, elliptic, deep green.* **Margins** *irregularly doubly serrate*

**Leaves** *have 12–15 pairs of veins, hairy below.* **Petiole** *2 cm, grooved*

**Fruit** *4 cm, erect, sub-sessile, with spreading scales*

**Bark** *bronze, yellowish or gray, peeling, aromatic, inflammable even when wet*

**Apex** *acute.* **Shoot** *dull gray-brown, initially pubescent*

Yellow birch is named after its vivid autumn colors, and grows to 15 m in its native eastern USA. It has an open or ovoid domed crown, and is notable for the long hairs on leaves and shoots.

**Flowers** open on ♂ and ♀ *catkins, before leaves, in late March.* ♂ **flowers** *yellow.* ♀ **flowers** 6 mm, *erect, red.*

**25 m. Crown** *broad, conic, often on 2 or 3 stems, purplish*

**♂ catkins** *5 cm when fully open*

**Catkins** *exposed over winter, purple*

**Bark** *fissured, gray, in small plates*

**Leaves** 10 cm, *persist until very end of autumn.* **Petiole** 3.5 cm, *speckled.* **Buds** 7 mm, *flat, on 3 mm stalks, narrowed at base, with 2–3 scales*

**Shoot** *purple by autumn*

**Fruit** 1.5 cm, *ovoid, set in clusters of 1–4*

**Leaves** *glabrous, paler below.* **Vein axils** *tufted*

**Seed** *small winged nutlet*

**Leaf margins** *slightly waved.* **Teeth** *irregular, shallow.* **Veins** *impressed*

**Fruit** *woody by time it ripens and releases seeds in October, very persistent as empty "cone"*

Alders grow in damp locations and have buoyant seeds that are distributed by water. The deep roots help to conserve river banks and improve the soil with their nitrogen-fixing nodules. The tree survives happily on drier sites but does not regenerate there. 'Laciniata' and 'Imperalis' are forms with deeply lobed leaves.

# Red alder

## *Alnus rubra*

**Fruit** a strobile (cone-like structure), 3 cm, ovoid or oblong, in clusters of up to 6 on orange peduncles to 15 cm, persist after opening. **Seeds** round

**Leaves** red-haired below. **Apex** acute

♀ **catkins** red, scaly, exposed over winter

**Leaves** 15 cm, ovate to elliptic. **Petiole** grooved

**Bud** 1 cm, dark red, stalked, scurfy. **Leaf base** rounded or cuneate

**Bark** gray, rough, inner bark bright orange-brown

**Bark** bears pale lenticels

**Shoot** at first green with long hairs, becoming bright shiny red, angled

With a wide range on the west coast, Red alder forms a domed tree to 25 m which is a prodigious colonizer of bare ground, preceding longer-lived conifers. White alder (*A. rhombifolia*) grows farther inland and has singly, not doubly, serrated leaves.

# Filbert

## *Corylus maxima*

**Leaves** 13 cm, cordate, broad obovate or rounded. **Margins** coarsely double-serrate. **Petiole** 1 cm, stout, pubescent, glandular

**Apex** acuminate

**Leaves** very hairy on undersides

**Fruit** subtended by incised leafy involucre or bract

**Shoot** with brownish, bristly, glandular hairs, usually persisting over winter

**Fruit** bract twice as long as nut

**Fruit** a 2 cm pointed nut, shiny, ovate-oblong, with pale hilium. **Bract** green, ripens brown

Filbert, a large shrub or small tree native to southeastern Europe, is planted in North America for its sweet oily nuts. Similar native species include the shrubby but occasionally tree-like Beaked hazel (*C. cornuta*), which has finely toothed leaves and an involucre constricted beyond the nut.

# European hazel

## Corylus avellana

♂ *catkins* 5 cm, *open Feb.* ♀ *catkins* 5 mm

*Shoot* covered in long stiff hairs with swollen tips.
*Leaf* 10 cm, hairy above and below doubly serrate

*Nuts* 2 cm,
ripen brown
in clusters
of 1–4

**10 m. Crown** shrubby

*Nuts* in papery, toothed bracts

European hazel is a shrubby tree native to almost all Europe and recognized by its nuts, densely haired twigs and leaves and its plump male catkins. Its female catkins are small and have red styles. 'Contorta', with twisted stems, is a useful ornamental.

# Hop hornbeam

## Ostrya virginiana

*Shoot* light green, hairy at first, orange
and lustrous by first summer, dark brown
when mature. *Bud* 6 mm, ovoid, acute

*Leaves* 15 cm, oblong-lanceolate,
yellow-green, paler below;
underside very hairy
at first, later
tufts in axils.
*Margins* doubly
serrate, teeth
incurved.
*Veins*
11–15
pairs

*Bark* thick, reddish brown,
broken into narrow adpressed
plates, shaggy at ends

♂ *catkins*
exposed over
winter

*Fruit* a winged nutlet
to 1 cm, in brown hop-
like bladder, in 7 cm
clusters

Hop hornbeam is a native of the eastern USA and forms a round-topped tree to 20 m. Ironwood (*O. knowltonii*), a rare tree from the southwest, has smaller 5 cm leaves with 5–8 pairs of veins. The seed, totally enclosed by the involucre, distinguishes *Ostrya* from *Carpinus*.

# American hornbeam

## *Carpinus caroliniana*

**Leaves** 10 cm, oval or ovate, long pointed, dull green above, light yellow-green below. **Base** rounded or cuneate, usually oblique. **Veins** slender, deeply impressed. **Petiole** 1 cm, hairy, bright red when young

**Leaf** glabrous above, hairy in vein axils below

**Margins** doubly serrate. **Teeth** sharp, spreading, glandular

**Shoot** slender, initially with long silky white hairs, becoming red-brown, shiny, later dull gray, tinged red

**Fruit** in 8–11 pairs, a nutlet subtended by trident bract, hanging in 15 cm clusters on slender pubescent red-brown stems. **Bract** leafy, green, ripening to brown. **Involucre** 4 × 2.5 cm, coarsely toothed, often only on one margin

**Catkins** 4 cm, appear with leaves in spring

**Nutlet** 8 mm

♂ **catkins** overwinter in bud, open in spring. **Buds** 3 mm, ovoid, acute. **Scales** brown with white margins

**15 m. Crown** usually bushy, dense and flat-topped. **Branches** zig-zag, slender, wiry, pendulous at tips

**Bole** short, characteristically fluted

**Bark** smooth, broad horizontal patches

American hornbeam, native to eastern America from the Great Lakes as far south as Honduras, is sometimes called Bluebeech on account of the similarity of its bark to Beech. It is a small tree, often a shrub, and assumes rich autumnal colors. European hornbeam (*C. betulus*) is more tree-like in habit, reaching 25 m, and is often planted as an ornamental in North America. It is best distinguished by its winter buds, which are slender, spindle-shaped and 6 mm or more in length.

# Beech and oak family Fagaceae

The members of this large, mainly temperate family have simple and alternate leaves and fruit in the form of nuts. These are either fully enclosed by an involucre (or cupule) of fused bracts as in *Fagus* (pp 102–4), and *Castanea* (p 120), or merely supported by it as in *Quercus* (pp 105–19). Male and female flowers are carried in separate catkins on the same tree.

## European beech ③ 🌳 🕯 🏚 🌿

### *Fagus sylvatica*

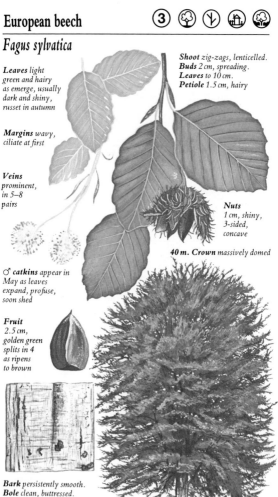

**Leaves** light green and hairy as emerge, usually dark and shiny, russet in autumn

**Margins** wavy, ciliate at first

**Veins** prominent, in 5–8 pairs

♂ **catkins** appear in May as leaves expand, profuse, soon shed

**Fruit** 2.5 cm, golden green splits in 4 as ripens to brown

**Shoot** zig-zags, lenticelled.
**Buds** 2 cm, spreading.
**Leaves** to 10 cm.
**Petiole** 1.5 cm, hairy

**Nuts** 1 cm, shiny, 3-sided, concave

*40 m.* **Crown** massively domed

**Bark** persistently smooth.
**Bole** clean, buttressed.
**Roots** often exposed by erosion

This majestic tree is instantly recognizable by its smooth bark. Although its natural range includes most of continental Europe and southern Britain, it has been widely planted elsewhere for its timber, and thrives on chalk and limestone. Oriental beech (*F. orientalis*), from the Balkans and Asia Minor, has leafy cupules and broader, larger leaves which have 7–10 pairs of veins.

# Some cultivars of Beech

**Copper beech** ('Purpurea') has strong purple pigments in leaves

**Copper beech leaves** unfold red or brownish pink but soon become a heavy shade of purple-green

**'Purpurea'** greener when grown in shade

**'Pendula'** and **'Dawyck'** both to 30 m tall

**Weeping beech** ('Pendula') has very pendent shoots on sprawling branches. **Leaves** wide. **Bark** is silver-gray at summit, where it is often exposed

**Dawyck beech** (F. sylvatica 'Dawyck') very fastigiate. **Branches** curving, upswept. Dense Lombardy poplar habit

**Petioles** 1.5 cm, pubescent

**Leaves** very deeply cut, hairy below

**Leaves** to 10 cm

**Fern-leaved beech** ('Heterophylla') has variable leaves, sometimes willow-like

Beech is noted for the variety of its cultivars. Copper beech, the most common, may have deep purple ('Purpurea') or red leaved clones. Weeping and Dawyck beeches have distinctive habits, 'Heterophylla' remarkable leaves. Others include 'Zlatia', with new foliage golden until July, and 'Rotundifolia' with small, almost round leaves and variegated forms.

## *Fagus grandifolia*

**Buds** 2.5 cm, slender, spindle-shaped, pale brown, set singly at sharp angle, made of many imbricate scales, finely hairy at apex

**Shoot** green initially, becoming reddish brown by winter, later ash-gray; has long hairs at first, soon lost

**40 m**, often surrounded by suckers. **Crown** rounded, compact, with spreading, slightly drooping branches. **Bark** smooth, gray

**Shoot** slender, often zig-zagged, spotted with oblong orange lenticels

**Petiole** 1.3 cm, hairy

**Leaves** 15 cm, oblong-obovate or elliptic, acuminate, with 11–15 pairs veins. **Base** cuneate

**Seeds** 2 cm, triangular, in pairs, shiny brown

♂ **catkins** a bunch of pendent stamens. **Cupule** woody, 4-valved, hairy, with recurved prickles

**Margins** coarsely serrate. **Teeth** spreading or incurved

**Fruit** on stout, hairy peduncle

American beech, native to the east from New Brunswick and the Great Lakes to the Gulf, is the only beech which produces suckers, and a thicket is often formed around the tree. The leaves are usually glabrous except along the midrib and on the underside veins, but downy forms are occasionally found. In the southern part of its range var. *caroliniana* is usually encountered, differing in its firmer, less coarsely toothed leaves, which are a darker, lustrous green above.

# Oaks *Quercus*

This large and important group of noble trees comprises over 400 separate species as well as many hybrids and is found throughout the temperate areas of Europe, Asia and North America. All oaks are noted for their acorns which may ripen over either one or two years and provide the best means of identification. Another is the arrangement of leaves and buds which cluster at the end of shoots and separate *Quercus* from most other genera. Oaks have separate male and female flowers appearing on the same tree; the females are erect but inconspicuous while the males appear in long, pendulous catkins at the same time as the new leaves open. The genus contains both deciduous and evergreen species.

## Sessile oak

### *Quercus petraea*

**40 m. Crown** tall

**Branches** straight, spreading, radiating. **Bole** long

**Buds** 6 mm, ovoid, with many finely pubescent scales, set in clusters at shoot tips, open in May. **Shoot** shiny, straighter than that of English oak

**Bark** thick, vertically fissured

**Leaves** 12 cm, hard or leathery, widest at middle, very flat, not so prone to caterpillar attack as English oak

**Leaf base** cuneate, without auricles

**Acorns** in clusters of 2–6, sessile or on very short peduncles

**Petiole** to 3 cm

**Leaf underside** has brownish pubescence on veins

**Lobes** shallow, in 5–9 pairs

With a long bole that extends well into the crown, this oak can be further distinguished from English oak by its almost stalkless, slightly smaller acorns and the wedge-shaped bases of its leaves. It grows well on light, acid, stony soil. The similar Downy oak (*Q. pubescens*) has densely hairy leaves and shoots.

## *Quercus robur*

**35 m. Crown** *wide*

*Foliage set in bunches*

*Branches* large and heavy, very irregular, bear sprouts

*Crown* irregularly domed. *Bole* short, very stout. *Bark* fissures into tiers of rectangular plates

*Lobes* in 4 or 5 pairs, deeply cut

*Leaves* have auricles at bases. *Leaf stalk* 1 cm

*Leaves* to 12 cm, widest above middle

*Acorns* 2.5 cm, usually in pairs on long, thin peduncle to 10 cm

*Cypress oak* (*Q. robur* 'Fastigiata') has upswept crown, narrow when young; grows in central Europe

English oak is very common throughout Europe and is also grown in the USA. It thrives on heavy clay but can adapt to the lighter, stonier soil favoured by *Q. petraea*. Unlike Sessile oak its leaves are glabrous below and have auricles at their bases. Its acorns are carried on long stalks and give the tree its alternative name of Pedunculate oak. Many predators such as the larvae of the Purple hairstreak butterfly (*Quercusia quercus*) attack its young foliage but do no permanent damage since further flushes of growth occur until September.

# White oak

## Quercus alba

**Leaves** 24 cm, obovate, open reddish, later bright green above, glaucous below. **Lobes** (7–9) rounded. **Sinuses** deep, rounded, nearly to midrib, variable between broad and narrow

**Base** narrowly cuneate. **Petiole** 2.5 cm, stout, glabrous, reddish

**Shoot** at first with hairs, red above, green below

**Bud** ovoid or globose. **Acorn** 2 cm, long ovoid

**30 m**, rarely to 45 m. **Crown** wide, irregularly domed, open, with heavy spreading limbs. **Bark** furrowed, scaly

**Acorns** sessile or on 5 cm peduncles, in clusters of 2–6, ⅓ enclosed by shallow, warty cupule

The most common of the white oaks, a group recognizable by their rounded lobes, this important lumber tree is found throughout the east and is recognizable by its leaves, the largest of any oak, and the shallow cupule of its acorn.

# Oregon white oak

## Quercus garryana

**Leaves** 15 × 5–12 cm, thick, leathery, hairy below. **Lobes** (5–9) may be slightly toothed

**Sinuses** very deep

**Acorn** 3 cm, ovoid, sessile or short-stalked. **Cupule** up to ⅓ encloses acorn, very thin, loose, scaly, hairy outside, inside glabrous, pale

**Apex** toothed or blunt. **Base** cuneate or rounded. **Margin** down-turned

**Petiole** 2.5 cm, stout, hairy

**Bud** 1.3 cm, acute, hairy

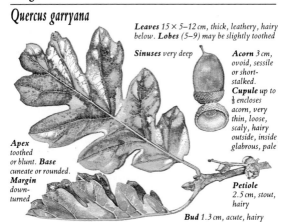

This species resembles the eastern White oak but is found along the western coastal strip from mid-California northward, being the only oak in Washington and British Columbia. It grows to 20 m with a broad crown, and the sweet acorns are edible.

# California white oak

## *Quercus lobata*

**30 m. Crown** open, wide, spreading, dividing low.
**Branches** drooping at tips. **Acorn** 5.5 cm, long conic

**Acorn**
single or in
pairs,
sessile, tip
rounded or
pointed

**Cupule** to
⅓ of
acorn,
hairy

**Bark** pale gray, thick,
scaly, in plates near
ground. **Leaves**
10 cm, thin, hairy
on both sides.
**Lobes** (7–11)
deep. **Sinuses**
narrow. **Petiole**
1.3 cm, stiffly
haired

**Cupule scales**
with free tips

The largest of the western oaks, this species, also known as
Valley oak, occasionally grows as high as 45 m, and can be
recognized by its unusual long-conic acorn. The buds are orange-
brown and hairy, and the acorn kernel is sweet.

# Post oak

## *Quercus stellata*

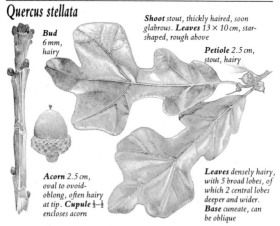

**Bud**
6 mm,
hairy

**Shoot** stout, thickly haired, soon
glabrous. **Leaves** 13 × 10 cm, star-
shaped, rough above

**Petiole** 2.5 cm,
stout, hairy

**Acorn** 2.5 cm,
oval to ovoid-
oblong, often hairy
at tip. **Cupule** ⅓–½
encloses acorn

**Leaves** densely hairy,
with 5 broad lobes, of
which 2 central lobes
deeper and wider.
**Base** cuneate, can
be oblique

Post oak rarely grows beyond 15 m, with a rounded crown, and
is distinguished from other white oaks by the rough upper surface
of its leaves. It is found throughout southern and eastern USA
and is an important lumber tree, producing a tough wood.

# Chestnut oak

## Quercus prinus

**Leaves** obovate or elliptic.
**Acorns** single or in pairs

**Shoot** stout, with pale red- or orange-brown hairs at first, later glabrous, ash-gray. **Acorn** 3.5 cm, shiny, on short, stout, hairy 1 cm stalk

**Leaves** 22 cm

30 m

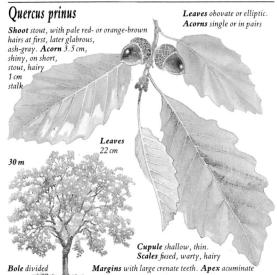

**Cupule** shallow, thin.
**Scales** fused, warty, hairy

**Bole** divided

**Margins** with large crenate teeth. **Apex** acuminate

Chestnut oak, from New England, is distinctive in its foliage, which resembles that of Chestnut (p120) although without the spiny tips to the lobes. Swamp chestnut oak (*Q. michauxii*), from the southeast, has obovate leaves, silvery white below, and cuneate scales on the thick cupule.

# Chinkapin oak

## Quercus muehlenbergii

**Shoot** slender, very hairy at first, red-brown, later glabrous, gray. **Bud** 4 mm, acute, ovoid. **Acorn** 2 cm, ovoid, chestnut brown or nearly black, sessile or short-stalked. **Cupule** up to ½ enclosing acorn

**Leaves** 18 cm, obovate or oblong-lanceolate, shiny yellow-green above, whitish, hairy below

**Petiole** 4 cm, slender

**Base** cuneate

**Apex** acuminate.
**Margins** regularly serrate

**Scales** free at tips

**Cupule** thin

**Teeth** incurved, ending in glandular tipped point

Growing in the east to 20 m, or rarely 50 m, this species is named after the resemblance of its foliage to the Chinkapins (p120). Swamp white oak (*Q. bicolor*) has broad obovate–oblong leaves and a 3 cm acorn on a 10 cm peduncle.

109

# Bur oak

## Quercus macrocarpa

**Shoot** *often develops corky wings*

**Leaves** *25 cm, obovate, thick, shiny above, gray-green and slightly downy below.* **Sinuses** *shallow, except central deep pair.* **Acorn** $\frac{1}{3}-\frac{1}{2}$ *enclosed by deep, mossy cupule.*

**Acorn** *to 5 cm, ellipsoid, with blunt or depressed apex*

**Lobes** *(5–9) irregular, rounded.* **Petiole** *2.5 cm, stout*

Bur oak is native to eastern central USA and down the Mississippi to the Gulf, and grows to 25 m, or rarely 50 m, with a broad, rounded crown. The acorns, with their distinctive cupules, are the largest of any native oak.

# Overcup oak

## Quercus lyrata

**Acorn** *2.5 cm, sessile or on 3 cm hairy stalk.* **Cupule** *usually encloses acorn, splitting 1st year*

**Leaves** *24 cm, irregular, usually narrow oblong*

**Lobes** *(5–9) rounded or short-pointed, longest at tip*

**Sinuses** *broad, rounded*

**Cupule** *thin, hairy, with fused ovoid scales.* **Acorn** *ovoid or spherical*

**Leaves** *hairy at first, later glabrous, although possibly slightly downy on paler underside.* **Shoot** *medium stout, initially pubescent.* **Bud** *3 mm, downy*

**Leaf base** *cuneate.* **Petiole** *2.5 cm*

**30 m. Crown** *rounded, open.* **Branches** *small, often droop.* **Bark** *gray, with thin scaly plates*

Overcup oak, which often makes no more than a comparatively small oak, is native to the southeastern USA and is easily identified by its cupule, which usually completely encloses the acorn. Like other white oaks the acorns are produced on the current year's shoots and ripen in the first year.

# Turkey oak

## *Quercus cerris*

**Stipules** *long,
2.5 cm, round all
buds.* **Shoot**
*pubescent*

**Bark
fissures**
*shallow;
deeper and
plating on
old trees*

**Acorn** *large,
2.5 cm, in sessile,
"mossy" cup
whose 4 mm
filaments are
parted, pointing
upwards on upper
half of cup*

**Leaves** *13 cm, very variable, rough,
glossy above, gray, downy below.*
**Lobes** *angular, usually in 5–9 pairs*

This hardy native of southern and south-west Europe naturalizes
easily and grows vigorously into a mature and massively domed
tree which can reach 40 m. It is more upright than *Q. robur* (p 106);
unlike all other oaks, it has stipules arranged round *all* its buds.

# Lucombe oak

## *Quercus hispanica* 'Lucombeana'

**Acorn** *large,
2.5 cm, in
shallow,
"mossy" cup
on 1 cm stalk.*
**Filaments**
*not parted
as* Q. cerris

**Stipules**
*surround
terminal
buds only*

**Leaves** *12 cm, almost
evergreen, mostly shed in
early spring, before new
growth.* **Blades** *glossy
above, lighter and finely
pubescent below.* **Petiole**
*1 cm.* **Shoot** *hairy*

**Lobes** *variable,
rounded, with
spines*

Lucombe oak is a form of the natural but variable hybrid of
Turkey oak and Cork oak (p 118) inheriting the habit, leaf shape,
stipulate buds and acorns of the former species and the bark and
semi-evergreen nature of the latter. This tree was first raised by
William Lucombe, an Exeter nurseryman, in about 1765.

111

## *Quercus canariensis*

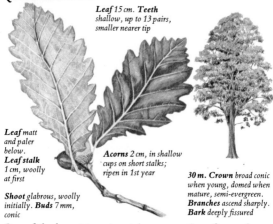

**Leaf** 15 cm. **Teeth** shallow, up to 13 pairs, smaller nearer tip

**Leaf** matt and paler below.
**Leaf stalk** 1 cm, woolly at first

**Acorns** 2 cm, in shallow cups on short stalks; ripen in 1st year

**Shoot** glabrous, woolly initially. **Buds** 7 mm, conic

**30 m. Crown** broad conic when young, domed when mature, semi-evergreen.
**Branches** ascend sharply.
**Bark** deeply fissured

One of the largest trees to retain a proportion of green leaves throughout the winter, this species is native to north Africa, Spain and Portugal. Caucasian oak (*Q. macranthera*) is similar but its wholly deciduous leaves have more rounded lobes and it carries dark red, shiny buds on pubescent shoots.

# Hungarian oak

## *Quercus frainetto*

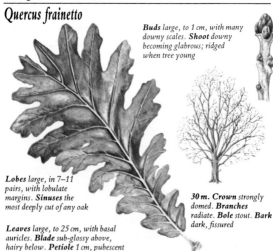

**Buds** large, to 1 cm, with many downy scales. **Shoot** downy becoming glabrous; ridged when tree young

**Lobes** large, in 7–11 pairs, with lobulate margins. **Sinuses** the most deeply cut of any oak

**Leaves** large, to 25 cm, with basal auricles. **Blade** sub-glossy above, hairy below. **Petiole** 1 cm, pubescent

**30 m. Crown** strongly domed. **Branches** radiate. **Bole** stout. **Bark** dark, fissured

Native to south-eastern Europe and planted as an ornamental in parklands in North America, this vigorous species is unmistakable in its boldly lobed foliage. Its acorns, similar to those of Algerian oak, ripen in one year, a characteristic of all the white oaks, and are carried at the end of the current year's shoot in a sessile cup with downy scales.

# Red oak

## *Quercus rubra*

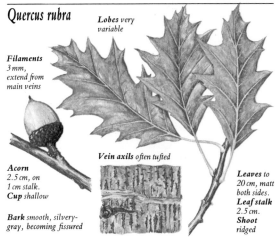

**Lobes** *very variable*

**Filaments** *3 mm, extend from main veins*

**Acorn** *2.5 cm, on 1 cm stalk.* **Cup** *shallow*

**Vein axils** *often tufted*

**Bark** *smooth, silvery-gray, becoming fissured*

**Leaves** *to 20 cm, matt both sides.* **Leaf stalk** *2.5 cm.* **Shoot** *ridged*

Glorious in the autumnal hues which give it its name, this sturdy species from the eastern USA typifies a large group of New World oaks which have filamented leaves and smoothish barks. Red oak has heavy branches and grows vigorously to produce a broadly domed crown reaching 35 m.

# Scarlet oak

## *Quercus coccinea*

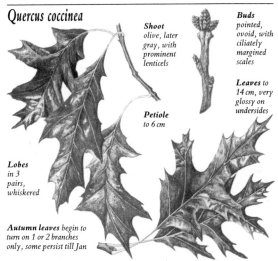

**Shoot** *olive, later gray, with prominent lenticels*

**Buds** *pointed, ovoid, with ciliately margined scales*

**Leaves** *to 14 cm, very glossy on undersides*

**Petiole** *to 6 cm*

**Lobes** *in 3 pairs, whiskered*

**Autumn leaves** *begin to turn on 1 or 2 branches only, some persist till Jan*

Scarlet oak has a more restricted range in the eastern USA than Red oak and assumes much richer autumnal colors, turning deep scarlet. For the rest of the year it is identifiable by its smaller, glossy, more deeply lobed leaves and smaller acorns. Its bud scales have pubescent margins and it grows to 25 m.

# Pin oak

## *Quercus palustris*

**New leaves** bright yellow, emerging with flowers in early May. **Buds** 3 mm

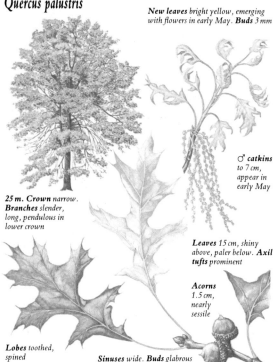

♂ **catkins** to 7 cm, appear in early May

**25 m. Crown** narrow. **Branches** slender, long, pendulous in lower crown

**Leaves** 15 cm, shiny above, paler below. **Axil tufts** prominent

**Acorns** 1.5 cm, nearly sessile

**Lobes** toothed, spined

**Sinuses** wide. **Buds** glabrous

Small pin-like branches which sometimes grow on its bole help to identify this bottom-land tree. Its prominent axillary tufts and its shallow acorn cups are also distinctive. Pin oak's narrowly lobed leaves turn red in autumn but not as richly as some of its cousins. Northern pin oak (*Q. ellipsoidsalis*) has sessile acorns to 2.5 cm.

# Black oak

## *Quercus velutina*

**Leaves** 23 cm

**Bud** 12 mm, angled, hairy. **Shoot** stout, densely hairy at first. **Leaves** hard, leathery. **Petiole** stout. **Leaves** have 5–7 lobes ending in bristly tips

**Acorn** 2 cm. **Cup** dull, scaly. **Acorn** ⅓–½ enclosed by involucre

**Leaves** coppery brown below, with reddish axillary tufts

**Leaves** shiny above

Widely distributed in the eastern USA, Black oak grows to 45 m with a narrow, open crown, and is distinguishable by the deep orange color of its inner bark showing in the fissures, its unusually tough leaves and its initially hairy shoots.

# Southern red oak

## *Quercus falcata*

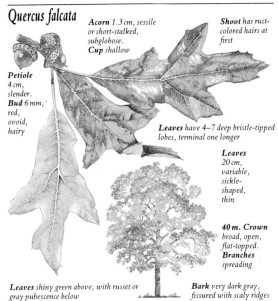

**Acorn** 1.3 cm, sessile or short-stalked, subglobose. **Cup** shallow

**Shoot** has rust-colored hairs at first

**Petiole** 4 cm, slender. **Bud** 6 mm, red, ovoid, hairy

**Leaves** have 4–7 deep bristle-tipped lobes, terminal one longer

**Leaves** 20 cm, variable, sickle-shaped, thin

**40 m. Crown** broad, open, flat-topped. **Branches** spreading

**Leaves** shiny green above, with russet or gray pubescence below

**Bark** very dark gray, fissured with scaly ridges

This variable species is native to the southeastern USA. The foliage illustrated above is typical, but sometimes leaves with three shallow lobes at the apex are found. The var. *pagodaefolia* has stout hairy petioles and leaves with 6–11 lobes.

# California black oak

## *Quercus kelloggii*

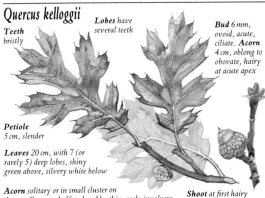

**Teeth** bristly

**Lobes** have several teeth

**Bud** 6 mm, ovoid, acute, ciliate. **Acorn** 4 cm, oblong to obovate, hairy at acute apex

**Petiole** 5 cm, slender

**Leaves** 20 cm, with 7 (or rarely 5) deep lobes, shiny green above, silvery white below

**Acorn** solitary or in small cluster on short stalk, up to half enclosed by thin, scaly involucre

**Shoot** at first hairy

Although similar to some eastern oaks, such as Black oak, this species is the only 'red oak' in the west and is easily identifiable in its native Oregon and California. It grows to 40 m with a broad, rounded crown and dark brown bark, becoming fissured with age and divided into oblong, scaly plates.

# Live oak

## *Quercus virginiana*

**Leaves** *12 cm, oblong-obovate or elliptic, shiny dark green above, pale, hairy below.* **Petiole** *1 cm, stout.* **Bud** *1.5 mm, globose*

**Acorn** *2.5 cm, 3–5 set on stout hairy peduncles to 12 cm, ellipsoidal, lustrous.* **Cupule** *⅓–½ acorn*

**Shoot** *slender, hairy*

**Base** *cuneate*

**Margins** *wavy or downturned, entire or remotely toothed.* **Bark** *rough, deeply furrowed*

*20 m.* **Crown** *very broad, to 50 m*

Live oak, so called because it is an evergreen, is native to the southeast coast and distinctive for its extraordinarily broad crown, which is often festooned with hanging clumps of Spanish moss. Its acorns ripen in the first autumn.

# Canyon live oak

## *Quercus chrysolepis*

**Apex** *acute*

**Leaves** *on young trees or vigorous shoots have spiny teeth*

**Flowers** *in racemes to 10 cm*

**Shoot** *thickly haired in 1st year.* **Bud** *acute, hairy*

**Leaves** *10 cm, elliptic or oblong-obovate, red hairs soon lost above, slowly below*

**Acorn** *5 cm, oval, tip hairy.* **Cupule** *thick, woolly*

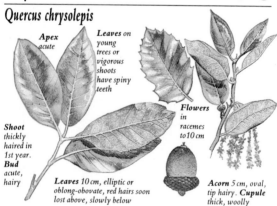

This southwestern oak reaches 30 m. Spined and entire leaves may be found on the same branch and persist 3–4 years. From California and with smaller acorns, Interior live oak (*Q. wislizenii*) has an acorn more than half enclosed in the cupule and keeps its leaves for two years while California live oak (*Q. agrifolia*) is one third enclosed and loses its leaves yearly.

# Willow oak

## *Quercus phellos*

**Leaf** 12 cm. **Petiole** 4 mm

**Acorn**
1 cm,
nearly
sessile

**Leaves**
entire

From the south-eastern USA, Willow oak has foliage which opens yellow and becomes golden in autumn. Schoch oak (*Q. x schochiana*) is its hybrid with Pin oak and has leaves with one or two lobes. Shingle oak (*Q. imbricaria*) has wider, entire leaves.

# Shingle oak

## *Quercus imbricaria*

**Leaves** 15 cm, elliptic or oblong-obovate, shiny above, pale, hairy below

**Apex**
cuneate.
**Margins**
often wavy

**Acorn**
2 cm, ovoid.
**Cupule** $\frac{1}{3}$–$\frac{1}{2}$
encloses acorn

**Scales**
hairy, adpressed

Shingle oak rarely grows much above 20 m tall, with a slender, rounded crown and a scaly, gray-brown bark with irregular shallow fissures. It is a native of the northeastern USA.

# Laurel oak

## *Quercus laurifolia*

**Acorn** 2 cm, ovoid or hemispherical, sessile or short-stalked, usually single. **Cupule** shallow, to $\frac{1}{4}$ of acorn. **Scales** pubescent

**Leaves** 10 cm, elliptic, very lustrous above, duller below, tapered at base and apex

**Leaves**
hairy at
first, lobed
irregularly.
**Leaves** flat or
wavy

**Shoot** slender, glabrous, with reddish tinge. **Bud** 7 mm, acute

Believed to be a natural hybrid between Willow oak and Water oak (*Q. nigra*) this rare tree grows to 20 m, or rarely 30 m, with a dense, rounded crown, along the Gulf and Atlantic coasts of the southeast, where it is often planted as a shade tree.

# Cork oak

## *Quercus suber*

**Leaves** *7 cm, hard, glossy above, gray felted below.* **Margins** *crinkled.* **Lobes** *very shallow, tipped with short, blunt spines.* **Petiole** *1 cm*

**Shoot** *short-haired.* **Buds** *2 mm, ovoid*

**Bark** *very thick with wide fissures and corky ridges*

**20 m. Crown** *open, domed.* **Branches** *heavy, twisted.* **Bole** *red-brown where bark removed*

The leaves of this Mediterranean tree resemble those of Holm and Turkey oaks (p 111). Its useful bark, harvested commercially about every ten years, is similar to that of Lucombe oak (p 111) and Chinese cork oak (*Q. variabilis*) whose wholly deciduous leaves have filamented teeth and are silvery-gray below.

# Holm oak

## *Quercus ilex*

**Young leaves** *yellow, hairy.* ♂ **flowers** *open in June*

**Leaves** *10 cm, variable*

**Acorns** *2 cm, pointed*

**Bark** *thin, in small plates, often curling*

**Leaves** *glossy, densely pubescent below*

**Spined leaves** *carried by young trees and sucker shoots*

A broadly domed species reaching up to 30 m, the somber Holm oak becomes brighter for a brief period in June when its new, yellowish leaves open. Kermes oak (*Q. coccifera*) is another native of the Mediterranean and hosts the Kermes insect from which scarlet grain dye is prepared. Its leaves are Holly-like.

# Blackjack oak

## *Quercus marilandica*

**Bud** *1 cm, rusty-haired*

**Acorn** *2 cm, ovoid-oblong, yellow-brown, often found in pairs, on short peduncle.*
**Cupule** *covers ⅓–⅔ acorn, thick, red-brown, hairy inside.* **Scales** *loosely overlapping, ciliate, hairy*

**Shoot** *stout, initially thickly haired, becoming ash-gray*

**Bud** *angled, conical.*
**Petiole** *2 cm, stout, yellow.*
**Leaves** *18 cm, obovate, as broad as long, with 3 (rarely 5) shallow lobes ending in a bristly tip at broad apex*

**Leaves** *shiny yellow-green above, scurfy, haired, below.*
**Margins** *entire or toothed.* **Base** *rounded or cuneate*

Remarkable for the shape of its leaves, this native of central and southeastern USA grows to 15 m with an irregular crown and rough, blackish bark furrowed into square plates.

# Roblé beech

## *Nothofagus obliqua*

**Margins** *irregularly serrate.* **Teeth** *sharp.* **Bud** *5 mm, ovoid, spreading*

**Leaves** *8 cm, ovate-oblong, deep green; pale, glaucous below.*
**Veins** *(7–11 pairs) impressed, extend to tips of teeth.*
**Base** *cuneate, slightly oblique.*
**Petiole** *5 mm, red*

**30 m.**
**Crown** *ovoid, open, spreading*

**Shoot** *slender, hairy, dark red above and paler below*

**Leaves** *sometimes develop lobes.* **Fruit** *8 cm, set singly, 4-ribbed*

**Fruit** *bears small scales*

**Lower branches** *pendulous.*
**Bark** *fissured*

The southern beeches are a genus of about 40 species native to the southern hemisphere and planted in North America only in gardens. They have smaller leaves and fruits than true beeches, and most are evergreen, although Roblé beech is not.

119

# American chestnut

## Castanea dentata

**Bud** ovoid. **Shoot** angled, shiny

**Leaf apex** acuminate

**Leaves** 20 cm, oblong-lanceolate, initially tomentose below, becoming glabrous. **Petiole** stout, angled

**Catkins** 12 cm, 1–3 female flowers at base of some male spikes, open after leaves

**Margins** have soft spiny glandular teeth

**30 m. Crown** low, broad, on a few large branches. **Bark** fissured into flat broad plates, often in spiral pattern

This once important species, native from New England down the Appalachians, has now been almost totally exterminated by Chestnut blight, a fungal disease. Chinese chestnut (*C. mollissima*), on which the fungus was introduced, is resistant, and is grown for its nuts. It has hairy buds and leaves with rounded bases.

# Golden chinkapin

## Chrysolepis chrysophylla

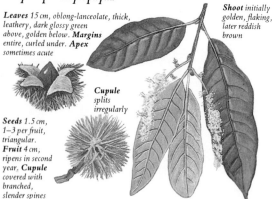

**Leaves** 15 cm, oblong-lanceolate, thick, leathery, dark glossy green above, golden below. **Margins** entire, curled under. **Apex** sometimes acute

**Shoot** initially golden, flaking, later reddish brown

**Cupule** splits irregularly

**Seeds** 1.5 cm, 1–3 per fruit, triangular. **Fruit** 4 cm, ripens in second year, **Cupule** covered with branched, slender spines

Very similar to Chestnut but with entire evergreen leaves, this species is native to the Pacific coast, where it grows to 30 m. Tanoak (*Lithocarpus densiflorus*) is similar in flower and range but has toothed leaves and an acorn fruit enclosed in a cup.

# Elm family Ulmaceae

This family of trees and shrubs has a worldwide distribution, and comprises about 15 genera, of which *Ulmus* and *Celtis* are most commonly encountered in North America. The family is characterized by simple, usually oblique, leaves that are carried alternately, and often have rough surfaces and buds with overlapping scales. Its flowers are usually perfect, have many stamens and appear with or before the leaves. The fruits are broad samaras in *Ulmus*, berries in *Celtis* and nutlets in *Zelkova*.

## American elm

### Ulmus americana

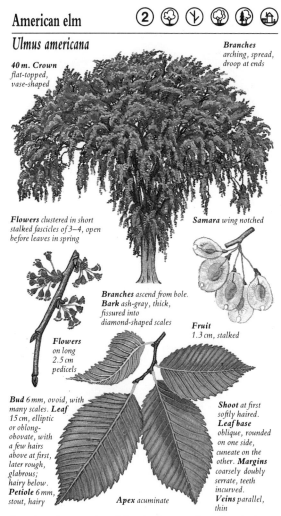

**40 m. Crown** *flat-topped, vase-shaped*

**Branches** *arching, spread, droop at ends*

**Flowers** *clustered in short stalked fascicles of 3–4, open before leaves in spring*

**Samara** *wing notched*

**Branches** *ascend from bole.* **Bark** *ash-gray, thick, fissured into diamond-shaped scales*

**Flowers** *on long 2.5 cm pedicels*

**Fruit** *1.3 cm, stalked*

**Bud** *6 mm, ovoid, with many scales.* **Leaf** *15 cm, elliptic or oblong-obovate, with a few hairs above at first, later rough, glabrous; hairy below.* **Petiole** *6 mm, stout, hairy*

**Apex** *acuminate*

**Shoot** *at first softly haired.* **Leaf base** *oblique, rounded on one side, cuneate on the other.* **Margins** *coarsely doubly serrate, teeth incurved.* **Veins** *parallel, thin*

American elm, a native of the east from southern Canada down to the Gulf, is recognizable by its crown and notched seed wing. It was once familiar in streets and gardens as a shade tree, but has been sadly depleted by Dutch elm disease.

# Rock elm

## *Ulmus thomasii*

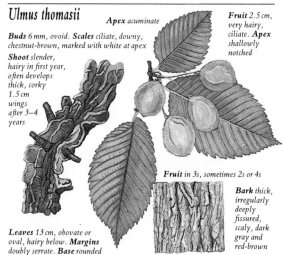

**Apex** *acuminate*

**Buds** *6 mm, ovoid. Scales ciliate, downy, chestnut-brown, marked with white at apex*

**Shoot** *slender, hairy in first year, often develops thick, corky 1.5 cm wings after 3–4 years*

**Fruit** *2.5 cm, very hairy, ciliate. Apex shallowly notched*

**Fruit** *in 3s, sometimes 2s or 4s*

**Bark** *thick, irregularly deeply fissured, scaly, dark gray and red-brown*

**Leaves** *13 cm, obovate or oval, hairy below. Margins doubly serrate. Base rounded*

Rock elm is native around and to the south of the Great Lakes, where it forms a tree to 30 m with a somewhat vase-shaped crown on a long straight bole. The very hairy buds and the flowers in racemes distinguish it from other elms. It produces a very tough and heavy timber, which is red-tinged.

# Slippery elm

## *Ulmus rubra*

**Leaf base** *oblique.* **Apex** *acuminate*

**Buds** *6 mm, ovoid obtuse, with long rusty hairs. Shoot hairy, ash-gray or orange-brown. Petiole hairy*

**Leaf margins** *doubly serrate. Teeth incurved.* **Bark** *shallowly fissured, scaly, dark brown and red; inner bark sticky*

**Leaves** *18 cm, scabrid with warts facing tip, downy below*

Named after the sticky aromatic substance which is found in its inner bark, Slippery elm is found throughout the eastern USA and forms a tree to 20 m. It can be distinguished from other elms by its leaves, which feel smooth if stroked in the right direction.

# English elm

## *Ulmus procera*

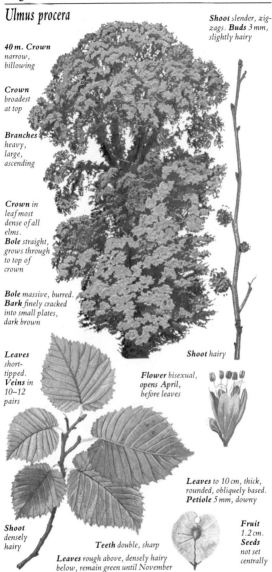

**Shoot** *slender, zig-zags.* **Buds** *3 mm, slightly hairy*

**40 m. Crown** *narrow, billowing*

**Crown** *broadest at top*

**Branches** *heavy, large, ascending*

**Crown** *in leaf most dense of all elms.* **Bole** *straight, grows through to top of crown*

**Bole** *massive, burred.* **Bark** *finely cracked into small plates, dark brown*

**Leaves** *short-tipped.* **Veins** *in 10–12 pairs*

**Shoot** *hairy*

**Flower** *bisexual, opens April, before leaves*

**Leaves** *to 10 cm, thick, rounded, obliquely based.* **Petiole** *5 mm, downy*

**Shoot** *densely hairy*

**Teeth** *double, sharp*

**Leaves** *rough above, densely hairy below, remain green until November*

**Fruit** *1.2 cm.* **Seeds** *not set centrally*

Until recently the most common English tree, this elm has been widely planted in the USA since the eighteenth century. Its origins, however, remain an enigma; it was either a very early introduction or may have arisen as a hybrid. Fertile seeds are rarely set and reproduction is nearly always by root suckering. Fluttering elm (*U. laevis*) has very long peduncles.

# Wych elm

## *Ulmus glabra*

**Leaves** large, 18 cm, variable

**Leaf tips** acuminate

**Shoot** stout, densely haired.
**Buds** large, 6 mm, hairy

**Seeds** central

**Bases** oblique.
**Petiole** 5 mm, usually hidden by leaf base

**Fruit** 1.5 cm

**Fruit** slightly notched, pale green in April when appears before leaves, shed in July when ripe

**40 m. Crown** broad, domed.
**Branches** arch. **Bole** short, stout, heavily buttressed, burred

**Bark** broadly ridged and cracked; silvery-gray, smooth, shiny on young trees. **Suckers** rare

Although sometimes shrubby on exposed sites, Wych elm is renowned for its majestic, spreading crown which is especially attractive in autumn. Its botanical name derives from the smoothness of its *young* bark and not from the leaves, which are very rough above. Most Wych elm seeds are fertile and it very rarely produces suckers. The fruit, appearing well before the foliage, is often produced so abundantly that the tree appears to be fully clothed. It is native to most of Europe and western Asia and the only elm unquestionably indigenous to Britain.

# Some cultivars of Wych elm

**Camperdown elm** *(left and below)*: **20 m. Crown** even

*Camperdown foliage* dense. *Leaves* large, 20 cm. *Branches* pendent, tortuous

*Camperdown branches* are grafted to Wych elm bole

**'Lutescen'** *leaves* large, 20 cm

**'Lutescens'** *leaves* pale in summer

**'Pendula'**: *12 m, asymmetrical, branches less pendent*

These compact, relatively small forms make ideal garden trees. Camperdown elm is the most common and has a more symmetrical crown than 'Pendula', whose branches form "herringbone" patterns. 'Lutescens' reaches 12 m.

# Siberian elm

## *Ulmus pumila*

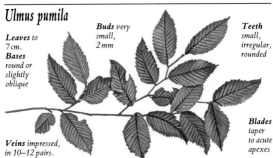

**Leaves** to 7 cm. **Bases** round or slightly oblique

**Buds** very small, 2 mm

**Teeth** small, irregular, rounded

**Veins** impressed, in 10–12 pairs. **Petiole** long, 5 mm

**Blades** taper to acute apexes

Almost evergreen and apparently very resistant to Dutch elm disease, Siberian elm grows fairly quickly into a small, flat-domed, leafy tree. Winged elm (*U. alata*), from the southeast, has corky, winged glabrous shoots and hairy fruits.

## *Ulmus carpinifolia*

**Seed** *1.5 cm, has closed apical notch*

**Leaves** *12 cm, shiny above, glabrous below*

*30 m.* **Crown** *variably domed.* **Branches** *ascend in arches.* **Branchlets** *long, pendulous.* **Bark** *fissured.* **Buds** *large, 5 mm*

**Veins** *have white axil tufts*

**Leaf** *base very oblique*

**Cornish elm:** *35 m.* **Crown** *vase-shaped.* **Branches** *form "fan" at apex.* **Foliage** *dense, carried above branches*

**Wheatley elm:** *35 m.* **Crown** *evenly conic.* **Branches** *set at c. 45°.* **Leaves** *(above) longer, rounder than Cornish elm*

Native to most of Europe, this hedgrow tree was probably introduced to southern Britain as a boundary marker in the first century BC and regional varieties — sometimes classified as species — have developed since then. Those most often seen are Wheatley (or Jersey) elm (*U.c.* var. *sarniensis*), planted in avenues, and Cornish elm (*U.c.* var. *cornubiensis*) which occurs "wild" in south-west England and southern Ireland. Its leaves are cupped and smaller than those of Wheatley elm.

# Chinese elm

## *Ulmus parvifolia*

**15 m.**
**Crown** *rounded dome, very dense*

**Leaves** *5 cm, oval, elliptic or obovate-lanceolate, lustrous above, glabrous except along midrib.* **Petiole** *8 mm, hairy*

**Branches** *pendulous with age*

**Margins** *singly serrate with forward crenate teeth*
**Base** *rounded, slightly oblique*

**Bark** *smooth, later flaking, cracked; gray or red-brown with large orange lenticels*

**Shoot** *very slender, hairy.* **Bud** *2 mm, ovoid*

**Apex** *acute*

Often semi-evergreen, this introduced ornamental produces its small greenish flowers in autumn at the base of the leaves, and is further distinguished by the herring-bone pattern of its shoots. Red elm (*U. serotina*) differs in its corky, winged shoots, its oblique, doubly serrate leaves and its flowers in racemes.

# Japanese zelkova

## *Zelkova serrata*

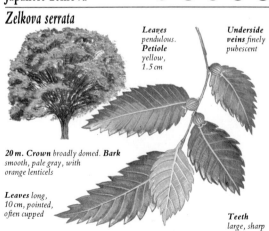

**Leaves** *pendulous.* **Petiole** *yellow, 1.5 cm*

**Underside veins** *finely pubescent*

**20 m. Crown** *broadly domed.* **Bark** *smooth, pale gray, with orange lenticels*

**Leaves** *long, 10 cm, pointed, often cupped*

**Teeth** *large, sharp*

Much prized in its native Japan for the quality of its timber, this zelkova is planted across the USA and is resistant to Dutch elm disease. Its larger leaves have pointed teeth and longer petioles and assume autumn shades of yellow, pink and bronze. Its fine zig-zagged shoots carry minute 1 mm buds.

# Hackberry

## *Celtis occidentalis*

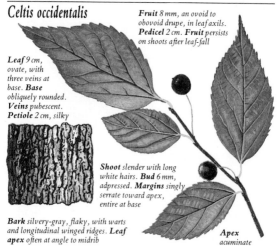

**Fruit** *8 mm, an ovoid to obovoid drupe, in leaf axils.* **Pedicel** *2 cm.* **Fruit** *persists on shoots after leaf-fall*

**Leaf** *9 cm, ovate, with three veins at base.* **Base** *obliquely rounded.* **Veins** *pubescent.* **Petiole** *2 cm, silky*

**Shoot** *slender with long white hairs.* **Bud** *6 mm, adpressed.* **Margins** *singly serrate toward apex, entire at base*

**Bark** *silvery-gray, flaky, with warts and longitudinal winged ridges.* **Leaf apex** *often at angle to midrib*

**Apex** *acuminate*

Hackberry, which grows to 20 m, or rarely 40 m, is one of a group of trees from the eastern USA similar to elms but differing in their fruit. Sugarberry (*C. laevigata*) has oblong-lanceolate leaves to 12 cm and orange-red fruits on shorter pedicels.

# Mulberry family Moraceae

## White mulberry

## *Morus alba*

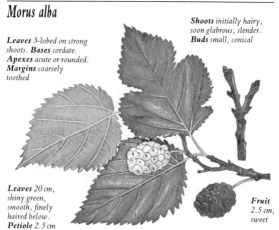

**Shoots** *initially hairy, soon glabrous, slender.* **Buds** *small, conical*

**Leaves** *3-lobed on strong shoots.* **Bases** *cordate.* **Apexes** *acute or rounded.* **Margins** *coarsely toothed*

**Leaves** *20 cm, shiny green, smooth, finely haired below.* **Petiole** *2.5 cm*

**Fruit** *2.5 cm, sweet*

White mulberry is a small Chinese tree reaching up to 16 m. Its fruit, which ripens from green to white, pinkish or violet purple, as insipid, although sweet. Black mulberry (*M. nigra*) has deeply cordate leaves, while those of Red mulberry (*M. rubra*) are truncate.

# Fig

## *Ficus carica*

**Leaves** 20 cm, usually deeply lobed but occasionally unlobed, cordate

**Leaves** shiny, white hairs on veins below. **Bark** smooth, patterned gray

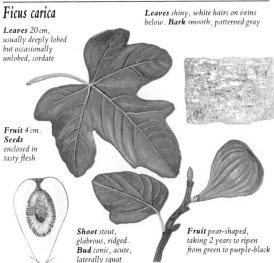

**Fruit** 4 cm. **Seeds** enclosed in tasty flesh

**Shoot** stout, glabrous, ridged. **Bud** conic, acute, laterally squat

**Fruit** pear-shaped, taking 2 years to ripen from green to purple-black

Fig, a native of southern Europe hardy only in warm climates, forms a spreading tree to 8 m, with a fruit developing from a modified shoot tip. Weeping fig (*F. benjamina*) has elliptic leathery leaves and dark red fruits. Fiddleleaf fig (*F. lysata*) has 50 cm glossy leaves with sinuate margins and white-spotted green fruit.

# Osage orange

## *Maclura pomifera*

**Shoot** zig-zags, bears 4 cm spine by axil

**15 m. Crown** rounded, open. **Bole** short, often crooked

**Petiole** 5 cm, hairy

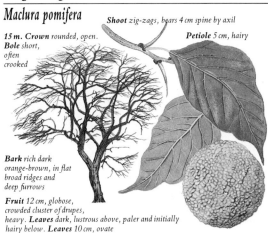

**Bark** rich dark orange-brown, in flat broad ridges and deep furrows

**Fruit** 12 cm, globose, crowded cluster of drupes, heavy. **Leaves** dark, lustrous above, paler and initially hairy below. **Leaves** 10 cm, ovate

The striking fruit of this widely planted windbreak is full of an acrid milky juice and small brown nutlets, and is totally unpalatable. The tree is often dioecious, although the flowers are inconspicuous, in either racemes (♂) or globose heads (♀).

# Katsura family Cercidiphyllaceae

## Katsura

## *Cercidiphyllum japonicum*

**Shoot** *very slender, reddish above, green below, sub-shiny, glabrous, thickened below buds*

**Margins** *serrate*

**Leaves** *8 cm, cordate, sometimes nearly circular, open bright pink, soon turning to fresh green; autumn color varies, may be gold, scarlet, purple*

**Leaf veins** *palmate on spur shoots, pinnate on long shoots, in 5s or 7s*

**Apex** *variable.*
**Petiole** *4 cm, red*

**Teeth** *shallow, rounded*

**Bud** *6 mm, shiny dark brown, set in opposite pairs*

**Spur shoots** *short, slow-growing.*
**Shoot** *straight, downcurved in outer crown*

Katsura is native to central China and Japan and is noted for its colorful foliage. It grows to 25 m with a columnar crown, often on a divided stem with dull gray fissured bark.

# Buckwheat family Polygonaceae

## Seagrape

## *Coccolobis uvifera*

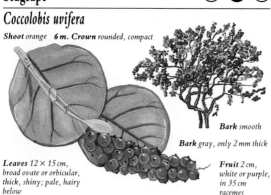

**Shoot** *orange*   **6 m. Crown** *rounded, compact*

**Bark** *smooth*

**Bark** *gray, only 2 mm thick*

**Leaves** *12 × 15 cm, broad ovate or orbicular, thick, shiny; pale, hairy below*

**Fruit** *2 cm, white or purple, in 35 cm racemes*

Seagrape is a small, open evergreen restricted to coastal sites in southern Florida. Its white flowers are followed by purple fruits.

# Papaya family Caricaceae

## Papaya

### *Carica papaya*

**Fruit** 12 cm, has thick skin and orange flesh

**Petiole** 90 cm, enlarged at base

**Petiole** hollow

**Fruit** yellow-green to orange. **Seeds** black

**Leaves** 60 cm

**Leaves** lobed palmately, toothed with pointed apexes

Papaya is a subtropical tree native from Florida southward which grows up to 10 m. It is cultivated for its fruits, which are carried on the stem of female or hermaphrodite trees.

# Magnolia family Magnoliaceae

## Saucer magnolia

### *Magnolia x soulangeana*

**Fruit** comprises spirally set carpels, each splitting to release 1 or 2 seeds

**Leaves** 15 cm (larger in some forms), downy below

**10 m. Crown** low

**Leaves** tapered at both base and apex, russet in autumn

**Flowers** to 20 cm. **Petals** (6–9) each 10 cm

**Bole** short. **Bark** smooth, gray. **Flowers** before or with new leaves

The classic magnolia, first raised in France, is a hybrid of two wild Chinese species, *M. denudata* and *M. lilliflora*. Also from China, Sprenger magnolia (*M. sprengeri*) has broader leaves and Sargent magnolia (*M. sargentiana*) obliquely cuneate ones.

# Southern magnolia

## *Magnolia grandiflora*

**Fruit** *10 cm, on stout 1.5 cm stalk, hairy*

**Petiole** *2.5 cm, stout, hairy*

**Flowers** *25 cm, from May to August*

**Leaves** *20 cm, thick, leathery, shiny, laurel-like, rufously pubescent below*

This small evergreen from the southern USA can be grown in less warm climates but may require the shelter of a wall to survive and flower. The Chinese Delavay magnolia (*M. delavayi*), also evergreen, has larger, very wide, matt, sea-green leaves.

# Japanese magnolia

## *Magnolia kobus*

**Bare branches** *regularly clothed in white flowers on older trees; younger trees shy to flower*

**Leaves** *15 cm.* **Shoot** *shiny, greenish-brown*

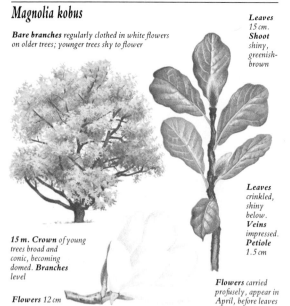

**Leaves** *crinkled, shiny below.* **Veins** *impressed.* **Petiole** *1.5 cm*

**15 m. Crown** *of young trees broad and conic, becoming domed.* **Branches** *level*

**Flowers** *12 cm*

**Flowers** *carried profusely, appear in April, before leaves*

This small species reaches 12 m and has slender, curved leaf buds and gray, downy flower buds. Its shoots are fragrant when crushed. *M. kobus* var. *borealis* is hardier and has pointed leaves. Those of Willow magnolia (*M. salicifolia*), also from Japan, are slender while its crown is an elegant, narrow dome.

# Cucumber tree

## *Magnolia acuminata*

**Flowers** 5 cm. **Petals** 8 cm, in 2 erect rings of 3, opening dull greenish yellow

**Leaves** to 23 cm, entire, fresh yellow-green above, pale, usually hairy below

**Margins** often wavy

**Leaf margins** entire

**Shoot** stout, glabrous. **Petiole** 3.5 cm, hairy

**Side buds** smaller

**Fruit** 8 cm, pink, later bright red, very erect, drops Sept

**Bud** to 1.5 cm, downy

From the eastern USA and named after the resemblance of its young seed clusters to cucumbers, this stately tree grows to about 25 m. Wilson magnolia (*M. wilsonii*), from China, hardly reaches a third of this and produces hanging, pure white flowers.

# Sweet bay

## *Magnolia virginiana*

**Fruit** 5 cm, comprising 6 mm flattened seeds. **Leaves** bluntly tipped with prominent midribs

**Flowers** to 7.5 cm, fragrant, June–July on leafy spur shoots

**Leaves** to 15 cm, shiny above

**Petioles** 2 cm, slender. **Shoot** lenticelled

**Leaf underside** glaucous, pubescent, minutely warty

Sweet bay is native to the coastal swamps and rivers of the eastern USA and evergreen in the southern part of this range. In the late seventeenth century it was the first magnolia grown in Europe where its small flowers sometimes persist till December.

# Tulip tree

## *Liriodendron tulipifera*

**Buds** 1 cm, flat, obovoid with 2 purple scales, stalked. **Shoot** shiny. **Leaf scars** very prominent

**Leaves** 15 cm, usually 4-lobed, glabrous, shiny, glaucous below

**Petiole** 10 cm, has 2 prominent stipules

**35 m. Crown** dense, becomes broader, more open with age. **Branches** regular

Native to eastern USA, this species, also called Yellow poplar, produces yellow-green tulip-shaped flowers in June, followed by erect, cone-shaped fruits. Chinese tulip tree (*L. chinense*) has thinner leaves which are deeply lobed and minutely warted below.

# Laurel family Lauraceae

## Avocado

## *Persea americana*

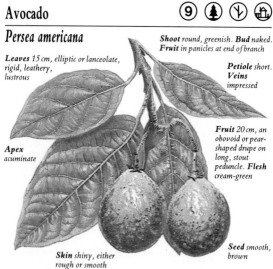

**Leaves** 15 cm, elliptic or lanceolate, rigid, leathery, lustrous

**Shoot** round, greenish. **Bud** naked. **Fruit** in panicles at end of branch

**Petiole** short. **Veins** impressed

**Apex** acuminate

**Fruit** 20 cm, an obovoid or pear-shaped drupe on long, stout peduncle. **Flesh** cream-green

**Skin** shiny, either rough or smooth

**Seed** smooth, brown

Growing to 20 m with a rounded, spreading crown, Avocado is widely planted in warm climates beyond its native Central America for its well-known fruit. It bears small greenish flowers.

# California laurel

## *Umbellularia californica*

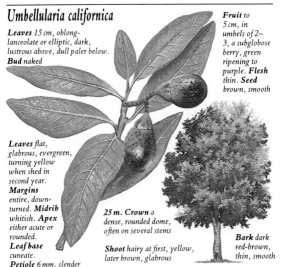

**Leaves** 15 cm, oblong-lanceolate or elliptic, dark, lustrous above, dull paler below. **Bud** naked

**Fruit** to 5 cm, in umbels of 2–3, a subglobose berry, green ripening to purple. **Flesh** thin. **Seed** brown, smooth

**Leaves** flat, glabrous, evergreen, turning yellow when shed in second year. **Margins** entire, down-turned. **Midrib** whitish. **Apex** either acute or rounded. **Leaf base** cuneate. **Petiole** 6 mm, slender

**25 m. Crown** a dense, rounded dome, often on several stems

**Shoot** hairy at first, yellow, later brown, glabrous

**Bark** dark red-brown, thin, smooth

This is the only species in this genus, named after its pale yellow flowers which appear in early spring in small umbels; it is also known as 'Headache tree' because of the unpleasant effect of too much sniffing of its aromatic foliage.

# Sassafras

## *Sassafras albidum*

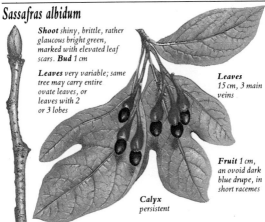

**Shoot** shiny, brittle, rather glaucous bright green, marked with elevated leaf scars. **Bud** 1 cm

**Leaves** very variable; same tree may carry entire ovate leaves, or leaves with 2 or 3 lobes

**Leaves** 15 cm, 3 main veins

**Fruit** 1 cm, an ovoid dark blue drupe, in short racemes

**Calyx** persistent

**Fruit** on 5 cm pedicel, thickened, red. **Leaves** thin, light green above, glaucous below. **Petiole** 3 cm

This freely suckering tree grows to 25 m with an open, flat-topped crown and closely furrowed dark brown bark. It is a native of the eastern USA, and the aromatic twigs, bark and roots have long been used to produce a fragrant oil or tea.

# Witch-hazel family Hamamelidaceae

## Sweet gum

### *Liquidambar styraciflua*

**30 m. Crown** *broad dome, ovoid-conic when young, turns orange, red or purple in autumn*

**Leaves** *5-rarely 7-lobed, 15 cm, glabrous except axillary tufts*

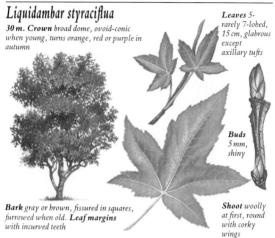

**Buds** *5 mm, shiny*

**Bark** *gray or brown, fissured in squares, furrowed when old.* **Leaf margins** *with incurved teeth*

**Shoot** *woolly at first, round with corky wings*

Sweet gum is native to the eastern USA and as far south as Guatemala. The three species in this genus have globose flowers and spiny pendulous fruits, like planetree. Leaves are similar to maples (pp 171–9) but alternate.

## Persian ironwood

### *Parrotia persica*

**Flowers** *before leaves*     **Shoot** *green-brown, has star-shaped stiff hairs*

**Leaves** *with 6–9 pairs impressed veins, brown hairs below*

**Buds** *downy, brown or black-purple*

**Flowers** *only have colored stamens*

**Bark** *similar to that of London plane, peels to expose brown or yellow*

**Leaf margins** *waved, glossy*

**Bark** *smooth, pink-brown or gray-green, exfoliating in large flakes*

Persian ironwood, native to the lush forests around the Caspian sea, can grow to 15 m, but is usually a sprawling shrubby tree. It is attractive in autumn when it assumes gold and crimson tints, and is distinguishable by its unusual bark.

# Rose family Rosaceae

The rose family is a large assemblage of trees, shrubs and herbs, with alternate leaves. The family is characterized by the flowers, which have the sepals, petals and stamens attached to the receptacle margin. The ovary may have one or several carpels, and may be either superior, above the petals and stamens, or inferior, below them. Four subfamilies are distinguished by fruit.

Two subfamilies, centered on *Rosa*, the Rose, and *Spirea*, contain only shrubs. *Prunoideae* is defined by having a fruit which is a drupe, a fleshy outer covering around a single bony seed. *Prunus* is the only tree genus, divided into sections based on the flower arrangements.

The subfamily *Pomoideae* has a fruit which is a pome, which has 2–5 carpels containing the seeds within a fleshy covering, and includes many tree genera.

# Downy serviceberry

## *Amelanchier laevis*

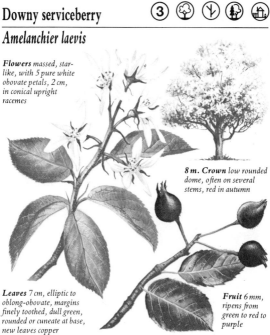

**Flowers** *massed, star-like, with 5 pure white obovate petals, 2 cm, in conical upright racemes*

**8 m. Crown** *low rounded dome, often on several stems, red in autumn*

**Leaves** *7 cm, elliptic to oblong-obovate, margins finely toothed, dull green, rounded or cuneate at base, new leaves copper*

**Fruit** *6 mm, ripens from green to red to purple*

Snowy mespil is a small tree or shrub, now naturalized on sandy heaths in Europe, native to eastern North America. The sweet black fruits are edible when they ripen in July. *Amelanchier* is a genus of small trees from North America and Asia.

# Single-seed hawthorn

## *Crataegus monogyna*

**Leaves** 10 cm, deeply lobed, with few teeth. **Thorns** to 2.5 cm

**Veins** pinnate. **Flowers** in corymbs of up to 16, appear with leaves in May

15 m. **Crown** dense. **Bole** short, fluted

**Shoot** purplish, stout. **Buds** 2 mm, glabrous, set in pairs at bases of spines

**Fruit** 1 cm, 1-seeded, has persistent calyx, profusely set, ripening September

Hawthorn's common names, "May" and "Quickthorn", derive from its flowering season and the speed with which it can form hedges. It has pink-flower cultivars such as 'Pendula Rosea'. Midland hawthorn (*C. oxyacantha*) is less spiny and has shallowly-lobed leaves, two-seeded haws and veins all pointing forward.

# Cockspur thorn

## *Crataegus crus-galli*

**Leaves** 8 cm

**Fruit** 1.5 cm, 2-seeded, sometimes persists in winter

**Leaves** serrate, glabrous. **Shoots and spines** purple-brown. **Spines** to 8 cm

Native to eastern USA, Cockspur thorn forms a low, rounded tree to 7 m and assumes rich orange tints in autumn. Frosted thorn (*C. pruinosa*) has elliptic, lobulate, coarsely serrated leaves to 4 cm; its 2 cm fruit is green with a glaucous bloom and has 5 seeds. Black thorn (*C. douglasii*) has broad obovate, serrate leaves to 10 cm and many shiny black 1.3 cm fruits.

# Mountain ashes Sorbus

This genus produces large corymbs of flowers developing into red, white, pink or russet heads of berries. While the leaves of mountain ashes, or rowans, are pinnate, those of the European whitebeams are simple with white undersides.

## Mountain ash

### Sorbus americana

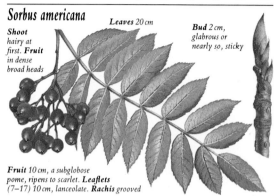

**Leaves** 20 cm

**Shoot** hairy at first. **Fruit** in dense broad heads

**Bud** 2 cm, glabrous or nearly so, sticky

**Fruit** 10 cm, a subglobose pome, ripens to scarlet. **Leaflets** (7–17) 10 cm, lanceolate. **Rachis** grooved

Native throughout northeastern North America from Newfoundland to the Appalachians, Mountain ash forms a tree to 10 m with thin, smooth or slightly scaly pale gray bark. It can grow on mountains and its leaves turn clear yellow in autumn.

## European mountain ash

### Sorbus aucuparia

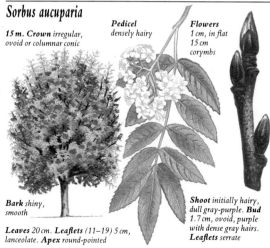

**15 m. Crown** irregular, ovoid or columnar conic

**Pedicel** densely hairy

**Flowers** 1 cm, in flat 15 cm corymbs

**Bark** shiny, smooth

**Leaves** 20 cm. **Leaflets** (11–19) 5 cm, lanceolate. **Apex** round-pointed

**Shoot** initially hairy, dull gray-purple. **Bud** 1.7 cm, ovoid, purple with dense gray hairs. **Leaflets** serrate

This native of Eurasia and North Africa has become naturalized in North America and is also widely grown as an ornamental. It can be distinguished from the native Mountain ash by its hairy buds and smaller leaflets with rounder apexes.

# Crab apple

## *Malus sylvestris*

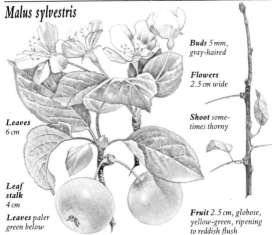

**Buds** *5 mm, gray-haired*

**Flowers** *2.5 cm wide*

**Shoot** *some-times thorny*

**Leaves** *6 cm*

**Leaf stalk** *4 cm*

**Leaves** *paler green below*

**Fruit** *2.5 cm, globose, yellow-green, ripening to reddish flush*

A native of Britain, western Europe and western Asia, Crab apple has white flowers, faintly flushed pink, and yields a fruit which, while hard and sour, makes excellent jelly. Crab apple is a parent of Orchard apple (*M. domestica*) which has much pinker flowers and sweeter, softer and larger fruit.

# Siberian crab

## *Malus baccata*

**15 m. Crown** *domed.* **Branches** *arch*

**Leaves** *to 7 cm, narrow, sharply serrate*

**Fruit** *1 cm, green, ripening to clear red.* **Pedicels** *up to 6 cm, very slender*

**Flowers** *to 4 cm wide, ivory white with spaced narrow petals*

**Petiole** *5 cm*

A tree with a wide natural distribution from Siberia through northern China to the Himalayas, this has fruit which persists through winter when the pointed buds also assist identification. Southern crab (*M. angustifolia*) has 5 cm leaves and 2.5 cm fruits.

# Purple crab

## *Malus x purpurea*

***Flowers*** *4 cm wide, open in early May in 6s and 7s.* ***Petals*** *obovate, darker in bud*

***Fruit*** *small, 2 cm*

The 8 cm leaves of this tree unfold purplish red in May but soon become green and glossier above with their main veins purplish below. The form is sparsely branched and reaches 6 m; its cultivars are bigger and more vigorous and include 'Aldenhamensis', 'Eleyi', 'Lemoinei' and 'Profusion'.

# Japanese crab

## *Malus floribunda*

***Teeth*** *coarse*

***10 m. Crown*** *dense.* ***Leaves*** *to 8 cm, paler and pubescent below, sometimes lobed on vigorous shoots*

***Flowers*** *4 cm, red in bud, fading to pink, then whitish, in umbels of 4–7*

***Fruit*** *2 cm, ripens yellow or red*

***Flowers*** *open in May*     ***Shoot*** *hairy, becoming almost glabrous*

Japanese crab is most attractive when displaying both its red buds and its pink and white flowers. Chinese crab (*M. spectabilis*) has broader, glossy leaves and large, pink, semi-double flowers. Neither species has apples with indented bases.

# Hupeh crab

## *Malus hupehensis*

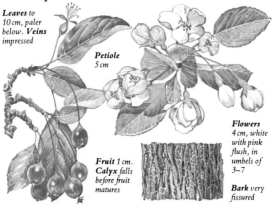

**Leaves** to 10 cm, paler below. **Veins** impressed

**Petiole** 5 cm

**Fruit 1 cm. Calyx** falls before fruit matures

**Flowers** 4 cm, white with pink flush, in umbels of 3–7

**Bark** very fissured

Hupeh crab has shiny, slightly downy, purple shoots and thorny spur shoots. The leaves can be made into a "red tea". It is very attractive in flower and vigorous, growing to 15 m. Sikkim crab (*M. sikkimensis*) has thornier spur shoots and woolly leaves.

# Chonosuki crab

## *Malus tchonoskii*

**Leaves** to 12 cm, opening silvery gray, tomentose

**Branches** long, very upright

**Crown** diamond-shaped

**Buds** rounded

**Shoot** has white wool, partly lost by autumn

**Fruit** 2 cm, yellow-green with purple flush

A native of Japan, this makes a good street tree because of its vigorous growth and upright habit. It has brilliant autumn colors. Yunnan crab (*M. yunnanensis*) has yellow-green leaves and 1–2 cm yellow or bright red fruit, speckled white.

# Wild pear

## *Pyrus communis*

**Petiole** *2–5 cm, slender*

**Teeth** *very fine*

**Flowers** *4 cm in corymbs 5–8 cm wide, appearing April, before leaves have fully emerged*

**Fruit** *4 cm, can be globose, has persistent calyx.* **Leaves** *8 cm, variable, glossy, paler below*

Wild pear is native to most of western Europe but most trees encountered have probably hybridized with selected orchard forms. The crown is narrowly conic at first, becoming tall and domed. 'Beech Hill' is spire-like. Callery pear (*P. calleryana*) has leathery leaves which are hairy on their midrib below.

# Willow-leafed pear

## *Pyrus salicifolia*

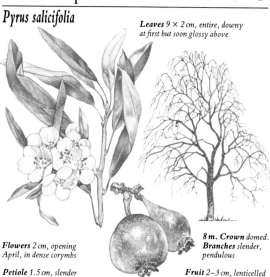

**Leaves** *9 × 2 cm, entire, downy at first but soon glossy above*

**Flowers** *2 cm, opening April, in dense corymbs*

**Petiole** *1.5 cm, slender*

**8 m. Crown** *domed.* **Branches** *slender, pendulous*

**Fruit** *2–3 cm, lenticelled*

A native of the Caucasus, this is usually seen as the 'Pendula' clone, which as a young tree has a distinctly weeping habit. Snow pear (*P. nivalis*), from southern Europe, has broader, less glossy leaves and rounded fruit. It reaches 20 m.

# Cherries and Plums *Prunus*

The trees in this genus, which also includes Blackthorn, Peach, Apricot and Almond, have fleshy single-seeded fruit which develops from a single ovary. Most species have up to four glands at the junction of their petioles and leaf bases; those without have shoots which are partially green for two or more years.

## Sweet cherry

### *Prunus avium*

**30 m. Crown** *open, conic when young*

**Branches** *whorled.* **Bole** *straight.* **Bark** *smooth with horizontal lenticels, fissures with age.* **Leaves** *ovate or oblong-ovate*

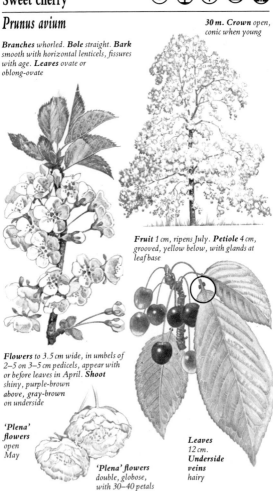

**Fruit** *1 cm, ripens July.* **Petiole** *4 cm, grooved, yellow below, with glands at leaf base*

**Flowers** *to 3.5 cm wide, in umbels of 2–5 on 3–5 cm pedicels, appear with or before leaves in April.* **Shoot** *shiny, purple-brown above, gray-brown on underside*

**'Plena' flowers** *open May*

**'Plena' flowers** *double, globose, with 30–40 petals*

**Leaves** *12 cm.* **Underside** *veins hairy*

Sweet cherry, also called Gean or Mazzard, is native to Europe and western Asia and is striking in flower, when its leaves are bronze, and also in autumn when they turn yellow and red. It is cultivated in areas too cold for apples and is the predominant parent of most domestic fruiting cherries. 'Plena' does not set fruit but its larger flowers persist for up to three weeks. Sour cherry (*P. cerasus*) has a shrubbier, suckering habit, glabrous leaves and tart fruit.

# Sargent cherry

## *Prunus sargentii*

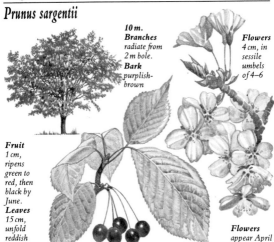

**10 m.
Branches** radiate from 2 m bole.
**Bark** purplish-brown

**Flowers** 4 cm, in sessile umbels of 4–6

**Fruit**
1 cm, ripens green to red, then black by June.
**Leaves**
15 cm, unfold reddish

**Flowers** appear April

Notable for the brilliant reds and scarlets of its early autumn foliage, Sargent cherry is a large tree that reaches 25 m when growing wild in the mountains of Japan. In cultivation, it is much smaller. It is identifiable by its dark lustrous bark.

# Autumn cherry

## *Prunus subhirtella* 'Autumnalis'

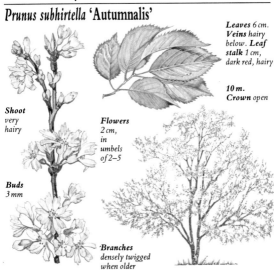

**Leaves** 6 cm.
**Veins** hairy below. **Leaf stalk** 1 cm, dark red, hairy

**10 m.
Crown** open

**Shoot** very hairy

**Flowers**
2 cm, in umbels of 2–5

**Buds**
3 mm

**Branches** densely twigged when older

This tree has the asset of flowering intermittently in winter in warm climates and its semi-double flowers are carried from October to early spring. It is a cultivar of the Japanese Higan cherry which is far less common and only carries flowers in April.

145

# Japanese cherries

Individually identifiable by their various flowers and habits, Japanese cherries comprise about forty small trees which have usually been grafted to a Wild cherry rootstock. Of mixed parentage, they are usually discussed as non-specific cultivars of *Prunus* rather than classified as varieties of *Prunus serrulata*, the species from which most have probably been developed. Collectively, they are sometimes referred to as the "Sato Zakura" (domestic cherries).

### 'Amanogawa'

*Flowers 5 cm wide in umbels of 3–6*

**Flowers** semi-double

**Petals** 6 mm

*Flowers densely clustered, open in May with bronze-green leaves*

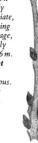

**8 m. Crown** tightly fastigiate, splaying with age, usually after 6 m. **Shoot** stout, glabrous. **Buds** large, 9 mm

### 'Kwanzan'

**13 m. Crown** usually an "inverted cone" (left). **Branches** spaced, cross and ascend at 45°, then spread with age (below)

*Flowers 5.5 cm wide, in umbels of 3–7. Pedicels 3 cm. Petals 20–30, open in April with bronze new leaves*

### 'Shimidsu'

**3 m. Crown** low, spreads

**Branches** often pendent

*Flowers 6 cm, pink in bud. Peduncle 6 cm*

All Sato Zakura have large leaves up to 20 cm long which turn gold or pink in autumn and shoots as shown. The 'Amanogawa' habit instantly identifies it, as do the ascending and crossing branches of the very common 'Kwanzan'; no other cherry is so sprawling in old age. The flowers of 'Shimidsu' are notable for their long penduncles.

**'Shirofugen'**

*Flowers* 5 cm, in umbels of 2–4

*Petals* 20–30

**6 m.**
**Branches** arch

**'Shirotae'**

*Flowers* 2 cm,
semi-double

**9 m. Branches** flat, then droop

*Flowers* droop

**'Tai-Haku'**

*Leaves* 20 cm, open golden coppery
**12 m**

*Flowers*
6 cm, largest
in genus, in 3s or 4s on 3 cm stalk

*Flowers*
hang outward

**'Ukon'**

*Branches* sparse, may arch on old trees

*Umbels* of
2–5 on stout stalk

*Flowers* 5 cm

*Leaves* open
bronze, soon
green

'Shirofugen', like 'Shimidsu', flowers very late, in mid-May, but has a vigorous, spreading habit, slightly larger leaves and flowers opening with a pink tinge. The semi-double flowers of 'Shirotae' open a month earlier. 'Ukon' is one of several Japanese cherries with yellow or greenish flowers. 'Tai-Haku' has the largest flowers of any *Prunus* species and is probably closest to the Wild cherry ancestor of the Sato Zakura.

# Japanese cherries

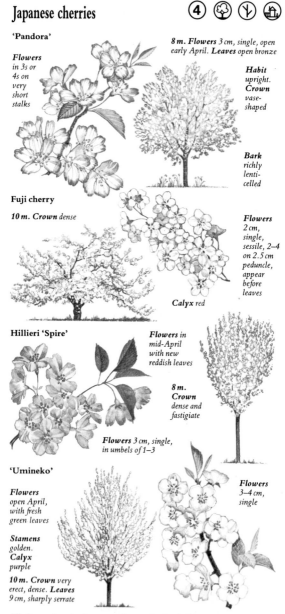

**'Pandora'**

*Flowers* in 3s or 4s on very short stalks

8 m. *Flowers* 3 cm, single, open early April. *Leaves* open bronze

*Habit* upright. *Crown* vase-shaped

*Bark* richly lenti-celled

**Fuji cherry**

10 m. *Crown* dense

*Flowers* 2 cm, single, sessile, 2–4 on 2.5 cm peduncle, appear before leaves

*Calyx* red

**Hillieri 'Spire'**

*Flowers* in mid-April with new reddish leaves

8 m. *Crown* dense and fastigiate

*Flowers* 3 cm, single, in umbels of 1–3

**'Umineko'**

*Flowers* open April, with fresh green leaves

*Stamens* golden. *Calyx* purple

10 m. *Crown* very erect, dense. *Leaves* 9 cm, sharply serrate

*Flowers* 3–4 cm, single

All these cherries have petals set singly and produce small black fruit. Fuji cherry (*P. incisa*) was crossed with Sargent cherry (p 145) to give *P.* x *hillieri*, whose 'Spire' clone is an ideal street tree, and with Oshima cherry (*P. speciosa*) to produce *P.* 'Umineko'. 'Pandora' is another non-specific *Prunus* cultivar (see p 146).

# Tibetan cherry

## *Prunus serrula*

**Flowers** 2 cm, in umbels of 2–4, on 4 cm stalks, appear in May. **Leaves** 12 cm, finely toothed

**Bark** glossy, with long bands of lenticels on vigorous trees, flaky

**Fruit** 5 mm, on 4 cm pedicels. **Leaves** hairy below. **Petiole** 1 cm

Primarily planted for the magnificence of its bark, Tibetan cherry is also noted for its finely serrate willow-like leaves. These are unique among true cherries. *P. x schmittii* has a similar but less spectacular bark, an upright habit and pink flowers.

# Yoshino cherry

## *Prunus x yedoensis*

**Leaves** to 15 cm, finely pubescent below. **Shoot** slender, hairy

**Flowers** 3.5 cm, set densely in umbels of 5–6, appear well before leaves, deep pink in bud, soon fade whitish

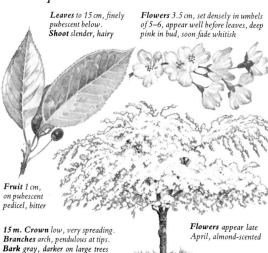

**Fruit** 1 cm, on pubescent pedicel, bitter

**15 m. Crown** low, very spreading. **Branches** arch, pendulous at tips. **Bark** gray, darker on large trees

**Flowers** appear late April, almond-scented

Yoshino is unknown in the wild but is believed to be a hybrid of Higan cherry (p 145) and Oshima cherry (*P. speciosa*). It is unusual among Japanese flowering cherries in having densely pubescent leaves, flower stalks and shoots. Its ripe fruit is black.

# Bird cherry

## *Prunus padus*

**Flowers** 1–1.5 cm, on crowded, irregular 15 cm racemes, open in late May, very fragrant

**Petiole** 2 cm, stout, grooved

**Fruit** globular

**Flower racemes** glabrous, semi-erect or drooping, leaves set at base

**Leaf veins** impressed

**Leaves** gray-green below, with axil tufts. **Teeth** fine

**Fruit** 8 mm, green, ripens red to black, bitter

**Leaves** to 12 cm, stout and leathery, dull

Bird cherry has a wide distribution in the USA and across northern Eurasia to Japan, and is easily recognized when in flower by its semi-erect racemes. Its smooth, dark, bitter-smelling bark was once used to prepare medicinal infusions. This tree has glabrous shoots and pointed and conic 5 mm buds.

# Black cherry

## *Prunus serotina*

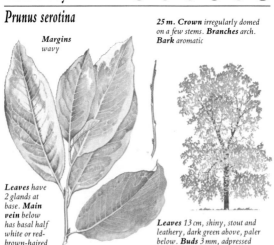

**25 m. Crown** irregularly domed on a few stems. **Branches** arch. **Bark** aromatic

**Margins** wavy

**Leaves** have 2 glands at base. **Main vein** below has basal half white or red-brown-haired

**Leaves** 13 cm, shiny, stout and leathery, dark green above, paler below. **Buds** 3 mm, adpressed

Black cherry, one of the largest trees in the genus, grows wild in eastern North America. Its fruits carry persistent calyxes, something missing from the fruit of Choke cherry (*P. virginiana*), which has thin leaves with spreading teeth. Pin cherry (*P. pensylvanica*) has glabrous leaves and flowers in umbels.

# Cherry laurel

## *Prunus laurocerasus*

**Flowers** in short racemes, appear in winter, open April.
**Fruit** 2 cm, ripens purple-black

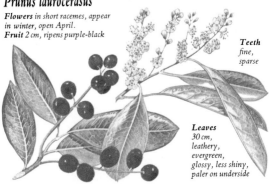

**Teeth** fine, sparse

**Leaves** 30 cm, leathery, evergreen, glossy, less shiny, paler on underside

From southeastern Europe, this species is often called "Laurel" since its foliage resembles that of Bay laurel (*Laurus nobilis*), used in wreaths and for cooking. Carolina cherry laurel (*L. caroliniana*) has oblong-lanceolate acuminate leaves to 12 cm.

# Portugal laurel

## *Prunus lusitanica*

**Leaves** evergreen, glossy above, dull below

**Petiole** 2.5 cm, grooved

**Fruit** 1.5 cm on 1 cm pedicels in stiff, pendent racemes, irregularly spaced

**Flowers** on dense 25 cm spikes

**Flowers** 1.2 cm, open mid-June

**Teeth** rounded

**Leaves** to 12 cm, leathery, glabrous.
**Peduncle** ridged behind pedicel

This evergreen species is native to the Iberian peninsula and while usually a shrubby tree of about 8 m may sometimes reach 16 m. Hollyleaf cherry (*P. ilicifolia*), from California, has ovate leaves to 5 cm with spiny teeth and almost sessile flowers.

# Cherry plum ● Purple-leaf plum

## *Prunus cerasifera* ● *P. cerasifera* 'Pissardi'

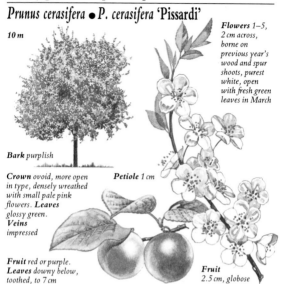

**10 m**

**Flowers** *1–5, 2 cm across, borne on previous year's wood and spur shoots, purest white, open with fresh green leaves in March*

**Bark** *purplish*

**Crown** *ovoid, more open in type, densely wreathed with small pale pink flowers.* **Leaves** *glossy green.* **Veins** *impressed*

**Petiole** *1 cm*

**Fruit** *red or purple.* **Leaves** *downy below, toothed, to 7 cm*

**Fruit** *2.5 cm, globose*

This attractive tree has an early flowering season, sometimes even in January, or as late as April. The normal green and purple leaf forms are about equally common. Willow-leaf plum (*P. salicina*), a native of China, is distinguishable by having its flowers in clusters of three and obovate-oblong leaves to 10 cm.

# Blackthorn

## *Prunus spinosa*

**Flowers** *single or in pairs, before leaves, 2 cm on a short creamy white 5 mm pedicel, open March–April*

**Leaves** *4 cm, sharply toothed, downy below with hairs on midrib vein.* **Petiole** *1 cm*

**Fruit** *1.5 cm, ripen blue to black, bloomed, on short pedicel*

**Shoot** *shiny gray, has terminal thorn.* **Buds** *small.* **Bark** *scaly, thorny*

**Leaves** *narrow, shape very variable from ovate to obovate, dull green*

Blackthorn or Sloe, native to Europe, is a suckering shrub or small tree to 6 m. In early spring it is a mass of small creamy-white flowers which obliterate the black of the bark and branches. The fruit ripens in October, and is used to flavour gin.

# Plum

## *Prunus domestica*

**Shoot** hairy when young, later slightly downy. **Fruit** black-purple or red

**Petiole**
2.5 cm

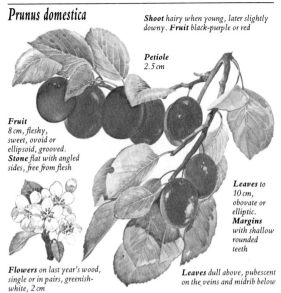

**Fruit**
8 cm, fleshy, sweet, ovoid or ellipsoid, grooved.
**Stone** flat with angled sides, free from flesh

**Leaves** to 10 cm, obovate or elliptic.
**Margins** with shallow rounded teeth

**Flowers** on last year's wood, single or in pairs, greenish-white, 2 cm

**Leaves** dull above, pubescent on the veins and midrib below

Plum, a small suckering tree to 10 m, is now thought to be a hybrid between Cherry plum and Blackthorn. Many named varieties are cultivated. American plum (*P. americana*) has flowers in umbels of two to five, and oblong-obovate leaves to 10 cm with an acuminate apex and doubly serrate margins.

# Apricot

## *Prunus armenica*

**Leaves** 10 cm, rounded or ovate, base cuneate, margins finely toothed

**Buds**
2 mm, single or in pairs

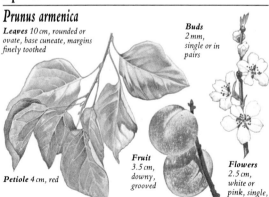

**Fruit**
3.5 cm, downy, grooved

**Flowers**
2.5 cm, white or pink, single, short-stalked

**Petiole** 4 cm, red

**Stone** smooth, free from flesh, margin thickened, ridged

Apricot makes a small rounded tree to 10 m, and is much cultivated for its delicious orange-red fruit. It does not come from Armenia but from northern China. Desert willow (*P. fremontii*), from the southwest, has 2 cm rounded or ovate leaves.

# Almond

## *Prunus dulcis*

**Flowers** appear in March, well before leaves, single or in pairs on a short stalk, bright pink; in older trees they appear only slightly in advance of leaves

**Leaf** usually carries glands at bottom of blade near petiole

**Petiole** 2.5 cm

**Leaves** lanceolate, 12 cm, glabrous, finely toothed margins and acuminate tip, "V"-folded near base and along midrib

**Shoot** green or purple. **Buds** 1–3 cm

**Fruit** 7 cm, velvety; hard thin fleshy coat splits to release seed

**Stone,** smooth, pitted

Almond grows wild, with spiny branches, in its native Mediterranean countries. Its early flowering season has endeared it to gardeners, although it is a short-lived tree with a spreading crown to 10 m. Its nuts are used in confectionery.

# Peach

## *Prunus persica*

**Leaves** do not have basal glands

**'Klara Meyer' flowers** pure pink, double, 4 cm, singly or in pairs. **Fruit** round, fleshy, 8 cm, stone deeply furrowed, clings to flesh

**Shoot** green or red-purple. **Shoot** glabrous, angular. **Buds** green and brown

**Leaves** 18 cm, narrow elliptic, finely serrate, matt green, paler below with raised midrib

Peach, which only grows to 8 m, is renowned for its juicy fruit, but is also notable for its single pale rose flowers when in blossom. It was introduced from China in the distant past. Nectarine differs only in having a glabrous fruit.

# Legume or Pea family Leguminosae

The Legume or Pea family, comprising several thousand species of trees, shrubs and herbs occurring all over the world, is characterized by pod-like fruits which have a line of seeds along the upper of the two seams. In most familiar species, which belong to the sub-family Lotoideae, the flowers consist of 10 stamens and 5 petals—a "standard" petal covering the top, two wing petals and two "keel" petals.

The second sub-family, the Caesalpinoidae has different flower forms, often with the front two petals enclosing all the others in bud. The Mimosoidae has tiny flowers clustered in heads or on racemes. All genera except *Cercis* have pinnate or bipinnate leaves.

## Judas tree

### *Cercis siliquastrum*

**15 m. Crown** one-sided.
**Bark** purplish, pinkish or gray, ridged at first, later becoming fissured

**Leaves** to 10 cm, either yellowish or dark green above and paler and glaucous beneath.
**Leaves** round with cordate base.
**Petiole** 5 cm

*Apex rounded or notched*

*Flowers* to 2 cm, in fascicles, on old or new shoots, branches, or even directly on bole

*Flowers* bright rose-purple; standard petal small, below wing petals

*Fruit* to 13 cm, thin, red-purple, ripening brown, flat, with 8–12 seeds, hairy at base

Judas tree, not, as commonly supposed, the tree upon which Judas Iscariot hung himself, but named after its native Judea, is unique in the Legume family for its orbicular leaves, similar to Katsura (p 130) but distinguishable by the paired buds. Eastern redbud (*C. canadensis*) differs in the shallowly cordate leaves, shiny above, with an abrupt acuminate apex and 1.2 cm flowers.

# Black locust

## *Robinia pseudoacacia*

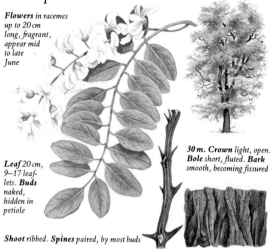

*Flowers in racemes up to 20 cm long, fragrant, appear mid to late June*

*30 m. Crown light, open. Bole short, fluted. Bark smooth, becoming fissured*

*Leaf 20 cm, 9–17 leaflets. Buds naked, hidden in petiole*

*Shoot ribbed. Spines paired, by most buds*

Black locust, also called False acacia, is native to the eastern USA. 'Frisia' has golden leaves. Clammy locust (*R. viscosa*) has glands on the shoot which secrete a sticky exudation.

# Honey locust

## *Gleditsia triacanthos*

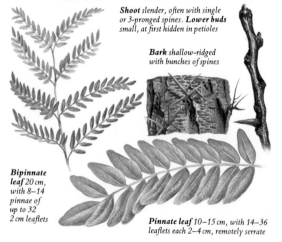

*Shoot slender, often with single or 3-pronged spines. Lower buds small, at first hidden in petioles*

*Bark shallow-ridged with bunches of spines*

*Bipinnate leaf 20 cm, with 8–14 pinnae of up to 32 2 cm leaflets*

*Pinnate leaf 10–15 cm, with 14–36 leaflets each 2–4 cm, remotely serrate*

Honey locust has a twisted pod 30–45 cm long, and thorns on the stem, although 'Inermis' is often preferred in towns as it lacks these. Water locust (*G. aquatica*) has oval 5 cm pods. Texas honey locust (*G. x texana*), with 12 cm pods, is their hybrid.

# Pagoda tree

## *Sophora japonica*

**25 m. Crown** uneven, low-spreading.
**Bark** has long, coarse ridges.
**Branches** very contorted

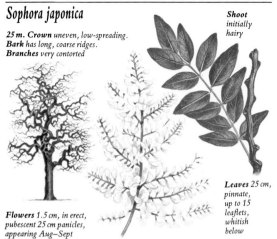

**Shoot**
initially
hairy

**Leaves** 25 cm,
pinnate,
up to 15
leaflets,
whitish
below

**Flowers** 1.5 cm, in erect,
pubescent 25 cm panicles,
appearing Aug–Sept

The Pagoda tree is a native of China and Korea and differs from Robinia in its pointed, hairy leaflets and its lack of spines. It produces white, pea-shaped flowers and pods of up to 8 cm. Texas sophora (*S. affinis*) has white flowers, tinged pink, in a raceme and pods constricted around each seed.

# Silver wattle

## *Acacia dealbata*

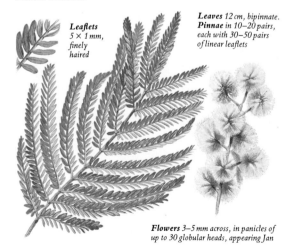

**Leaflets**
5 × 1 mm,
finely
haired

**Leaves** 12 cm, bipinnate.
**Pinnae** in 10–20 pairs,
each with 30–50 pairs
of linear leaflets

**Flowers** 3–5 mm across, in panicles of
up to 30 globular heads, appearing Jan

This short-lived Australian tree is common only in the warmest areas of California, where it is widely planted as a street tree. It rarely exceeds 15 m. Mimosa, or silk tree, (*Albizia julibrissin*) has larger, deciduous leaves and pink flowers.

# Mesquite

## *Prosopis juliflora*

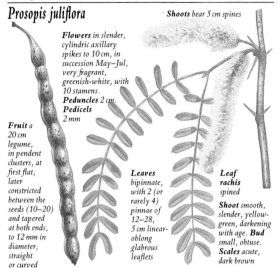

**Shoots** bear 5 cm spines

**Flowers** in slender, cylindric axillary spikes to 10 cm, in succession May–Jul, very fragrant, greenish-white, with 10 stamens. **Peduncles** 2 cm. **Pedicels** 2 mm

**Fruit** a 20 cm legume, in pendent clusters, at first flat, later constricted between the seeds (10–20) and tapered at both ends, to 12 mm in diameter, straight or curved

**Leaves** bipinnate, with 2 (or rarely 4) pinnae of 12–28, 5 cm linear-oblong glabrous leaflets

**Leaf rachis** spined

**Shoot** smooth, slender, yellow-green, darkening with age. **Bud** small, obtuse. **Scales** acute, dark brown

Mesquite, from the southwest, is a shrubby tree to 8 m, or rarely 15 m, often on several stems, with an open, irregular crown and thick, fissured dark red-brown bark. Screwbean mesquite (*P. pubescens*) differs in its thick, spirally twisted pod, which only reaches 5 cm, and the leaflets with 10–16 hairy segments.

# Royal poinciana

## *Delonix regia*

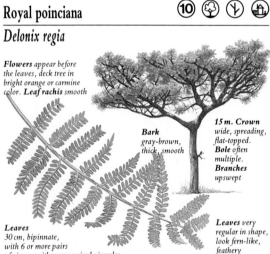

**Flowers** appear before the leaves, deck tree in bright orange or carmine color. **Leaf rachis** smooth

**Bark** gray-brown, thick, smooth

**15 m. Crown** wide, spreading, flat-topped. **Bole** often multiple. **Branches** upswept

**Leaves** 30 cm, bipinnate, with 6 or more pairs of pinnae with many paired pinnules

**Leaves** very regular in shape, look fern-like, feathery

Royal poinciana is a native of Madagascar but is now widely planted as an ornamental in tropical and subtropical regions, being valued for its dense racemes of showy flowers, with five petals with wavy margins. The fruit is a red-brown legume.

# Kentucky coffee tree

## *Gymnocladus dioicus*

**Shoot** *very stout, red downy at first, bloomed*

**Leaves** *large, to 1 m × 60 cm, bipinnate, with 4–9 pairs of pinnae, without terminal*

**Pinnae** *with 6–14 pinnules, 8 cm, stalked, entire, ovate*

**Leaflets** *at base (1–4) often simple*

**Leaflets** *open pink.* **Rachis** *persistent*
**30 m. Crown** *open, domed or conical.* **Bark** *flaking in reflexed scales*

With stout shoots and enormous leaves, the Kentucky coffee tree, from central eastern USA, is unmistakable. Separate trees bear male and female flowers, the latter in 30 cm panicles, and the 2 cm black seeds, in pods to 30 cm, were once used to make coffee.

# Yellow-wood

## *Cladrastis lutea*

**Leaves** *30 cm*

**Leaflets** *(5–11) 10 cm, obovate, widely spaced*

**Veins** *impressed*

**Leaflets** *usually alternate but variable.* **Buds** *naked, in 4s, hidden by base of petiole*

**Bark** *smooth, silvery gray or brown.* **Shoot** *slender*

**Shoot** *shiny*

This rare native of the mid-east forms a tree to 15 m with a rounded crown and very bright green foliage. The white flowers are in terminal panicles to 40 cm, and the fruit is a flat, acuminate 8 cm pod. *Cladrastis*, meaning brittle, refers to the shoots.

## *Laburnum* x *watereri* 'Vossii'

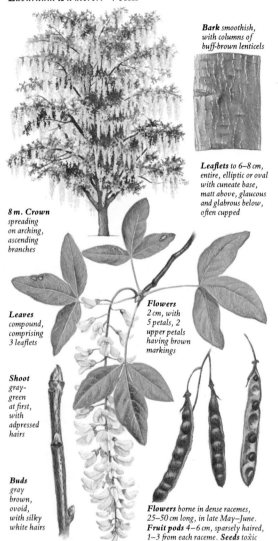

**Bark** *smoothish, with columns of buff-brown lenticels*

**Leaflets** *to 6–8 cm, entire, elliptic or oval with cuneate base, matt above, glaucous and glabrous below, often cupped*

**8 m. Crown** *spreading on arching, ascending branches*

**Leaves** *compound, comprising 3 leaflets*

**Flowers** *2 cm, with 5 petals, 2 upper petals having brown markings*

**Shoot** *gray-green at first, with adpressed hairs*

**Buds** *gray brown, ovoid, with silky white hairs*

**Flowers** *borne in dense racemes, 25–50 cm long, in late May–June.* **Fruit pods** *4–6 cm, sparsely haired, 1–3 from each raceme.* **Seeds** *toxic*

One of the showiest of small garden trees, this hybrid of Laburnum (*L. anagyroides*) and Scotch laburnum (*L. alpinum*) combines the longer flowers of the former species with the dense racemes of the latter; both differ from the hybrid in having silky hairs on their leaf undersides. Hop tree (*Ptelea trifoliata*), in the Rutaceae (p 165), has similarly compound leaves spotted with oil glands. Its seeds have flat wings like those of elms (pp 121–7).

# Quassia family Simaroubaceae

## Tree of heaven

### *Ailanthus altissima*

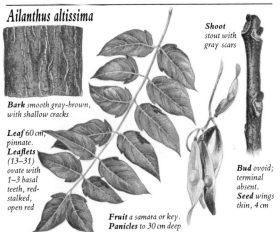

**Shoot** *stout with gray scars*

**Bark** *smooth gray-brown, with shallow cracks*

**Leaf** *60 cm, pinnate.* **Leaflets** *(13–31) ovate with 1–3 basal teeth, red-stalked, open red*

**Bud** *ovoid; terminal absent.* **Seed wings** *thin, 4 cm*

**Fruit** *a samara or key.* **Panicles** *to 30 cm deep*

Tree of heaven, with a domed crown to 25 meters, comes from north China. It grows rapidly and tolerates pollution, thriving in city streets and parks. Suckering prodigiously, it can produce leaves up to a meter long when such shoots are cut back.

# Box family Buxaceae

## Box

### *Buxus sempervirens*

**Flowers** *either ♂ or ♀, clustered in leaf axils*

**10 m. Crown** *often columnar, leaning.* **Bark** *cracked into square plates.* **Leaf margins** *rolled.* **Petiole** *1 mm*

**Leaves** *3 cm, leathery, yellowish below.* **Fruit** *1 cm, a capsule of 3 segments*

Native to some chalky areas of southern England, Box is more common in Mediterranean countries. Its green shoots are square in section and covered in orange hairs while its flowers open in April. It is an ideal tree for hedging and topiary since its small, evergreen leaves are able to withstand repeated clipping. Its hard, heavy wood is much sought-after for engraving.

# Mahogany family Meliaceae

## Chinaberry

### *Melia azedarach*

**15 m.** *Crown* rounded, dense and spreading. **Leaves** deciduous but semi-evergreen in extreme south of range

**Flowers** 2 cm, fragrant, in axillary panicles, with 5 mauve petals and narrow violet tube

**Bark** brown or red-brown, furrowed

**Leaves** 60 cm, bipinnate, about 5 pinnae

**Fruit** 1.3 cm, a yellow drupe

**Fruit** becoming wrinkled. **Flesh** thin, odorous

**Leaves** hairy, soon glabrous

**Leaflets** have 5–9 ovate or oval 5 cm pinnules. **Apex** acuminate. **Margins** serrate or entire and lobed

The fruit of this fast-growing but short-lived tree is poisonous, but its hard bony seeds are used as beads. It is a native of Indo-China now extensively planted in subtropical areas.

# Cashew family Anacardiaceae

## Cashew

### *Anacardium occidentale*

**Leaves** 10 cm, oval, leathery, entire, simple

**Fruit** 10 cm nut, kidney-shaped, in swollen fleshy stalk

**Flesh** sweet

**Leaf base** tapered

**Fruit** in complex inflorescences

**Flesh** ripens scarlet

The Cashew tree, which reaches 12 m, is widely grown not only for its familiar nuts but also for the 'Cashew apple', the edible flesh which develops from the flower stalk and which is used to make a lubricant. It is native to South America.

# Pistachio

## *Pistachio vera*

**Leaves** palmately compound, with 3–11 leaflets, on long petiole. **Leaflets** 6 cm, ovate

**Fruit** 2.5 cm, an ovoid drupe, in erect panicles to 10 cm, reddish, develops from flowers on ♀ trees only

**Stone** bony. **Seed** green or yellow

**Leaflets** sessile, at first downy, later leathery. **Apex** acuminate

Pistachio, a native of western Asia, grows to 10 m, and is valued for the nut kernels, much used in confectionery. Pepper tree (*Schinus molle*), a pendulous-branched tree to 15 m, has 15–25 toothed leaflets and pink pepper-like fruits.

# Staghorn sumac

## *Rhus typhina*

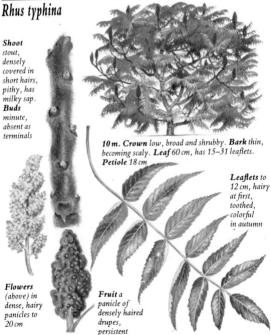

**Shoot** stout, densely covered in short hairs, pithy, has milky sap. **Buds** minute, absent as terminals

**10 m. Crown** low, broad and shrubby. **Bark** thin, becoming scaly. **Leaf** 60 cm, has 15–31 leaflets. **Petiole** 18 cm

**Leaflets** to 12 cm, hairy at first, toothed, colorful in autumn

**Flowers** (above) in dense, hairy panicles to 20 cm

**Fruit** a panicle of densely haired drupes, persistent

The brilliance of its autumn foliage and the curiosity of its crimson "lollipop" fruits persisting well into winter have made this sumac, native to the woods of the northeastern USA, a popular garden tree. Its common name refers to its hairy twigs.

# Holly family Aquifoliaceae

## American holly

### *Ilex opaca*

**15 m,** occasionally 30 m. **Crown** narrow, dense, conical. **Leaves** more spiny on lower branches within reach of animals. **Crown** of wild trees more open.

**Branches** upswept at tips

**Shoot** densely hairy at first, red-brown, later smooth, brown

**Leaves** 10 cm, thick, leathery, glossy, dark green above, yellowish below with prominent veins, persist 3 yrs

♂ **flowers** in cymes of 3–9, ♀ single or in 3s

**Bark** light gray, thin, smooth but warty.
**Bud** small, green, sharply pointed.
**Fruit** ripens autumn, persistent

**Fruit** 6 mm, ovoid, in clusters

**Skin** tough. **Flesh** yellow, holds 4 ribbed yellow seeds

A slow growing tree, this species is widely distributed in the southeast and as far north as Massachusetts, and is often cultivated and clipped in hedges. The berried branches of female trees are used for decoration. English holly (*I. aquifolium*) has smaller more glossy leaves and flowers on last year's shoots.

## Yaupon

### *Ilex vomitoria*

**Leaves** elliptic to oblong

**Bark** very thin, only 6 mm, light red-brown, developing small scales

**Margins** serrate. **Petiole** grooved

**Leaves** 5 cm. **Apex** obtuse

**Leaves** shiny dark green above, pale, opaque below. **Bud** obtuse, very dark

**Fruit** 6 mm, in great abundance on previous year's shoots, short-stalked, soon falls

Native to the southeast, Yaupon does not grow beyond 8 m but is noticeable because of the abundance of its flowers and berries. Dahoon (*I. cassine*), from the same region, has usually entire 7 cm leaves and shoots initially with silky hairs.

# Rue family Rutaceae

## Lime

### *Citrus aurantifolia*

*Flowers 2.5 cm, in racemes, with many stamens and 5 petals, white with purple margins.* **Leaves** *fragrant when crushed*

**Leaves** *9 cm, ovate.* **Petiole** *winged*

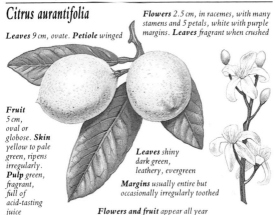

**Fruit** *5 cm, oval or globose.* **Skin** *yellow to pale green, ripens irregularly.* **Pulp** *green, fragrant, full of acid-tasting juice*

**Leaves** *shiny dark green, leathery, evergreen*

**Margins** *usually entire but occasionally irregularly toothed*

**Flowers and fruit** *appear all year*

The lime is a small, shrubby tree, usually only to 6 m, which often grows on several twisted stems. It is native to the Malay peninsula but is now widely cultivated in southern Florida, Arizona and California for its fruit, the heaviest crops being produced in summer. It is less hardy than other citrus trees.

## Lemon

### *Citrus limon*

**Leaves** *15 cm, elliptic, evergreen, glossy dark green above, paler green and dull below.* **Petiole** *slightly winged, with prominent joint at base where it joins the blade*

*6 m.* **Crown** *low, spreading, dense but open.* **Branches** *ascend.* **Shoot** *green, with stout spines*

**Margins** *very finely toothed*

**Bark** *light brown, knobbly*

**Fruit** *10 cm, green, ripening to yellow, elliptic, with nipple-like apex.* **Flesh** *greenish-yellow, juicy, bitter.* **Seed** *ovoid, pale brown*

**Leaves** *often crinkled.* **Apex** *bluntly short acuminate*

The lemon was introduced from the Far East, and although the fruit is a very familiar one the tree is less widely distributed in the south than the lime, which is tolerant of a wider range of soils. The tree fruits throughout the year, as well as carrying flowers which are very fragrant, white with a purple flush on the outside, and which have up to 40 stamens.

# Mandarin

## *Citrus reticulata*

**Shoots** *green, with fine thorns*

**Leaves** *10 cm, oval or lanceolate, glossy dark green above, paler and dull below, entire*

**Petiole** *1 cm, not winged*

**Fruit** *8 cm, globose with broad dimpled ends*

**Fruit** *divided into several fleshy segments inside loose, pithy skin.* **Seeds** *attached to central edge of segments*

Although it can tolerate only the mildest of frosts, Mandarin or Tangerine is one of the hardiest of fruiting citrus species, and is further distinguished by its smaller leaves, its unarmed petiole and the loose skin of its fruit. It is a native of China and grows to 6 m with a spreading crown.

# Sweet orange

## *Citrus sinensis*

**Shoot** *occasionally bears spines.* **Leaves** *15 cm, oblanceolate to ovate.* **Petiole** *narrowly winged*

**Flowers** *3 cm, white, fragrant, single or in small lax clusters*

**10 m. Crown** *rounded, dense.* **Bole** *short.* **Branches** *horizontal*

**Fruit** *10 cm, globose, orange-red or red, more rarely green or yellow, appear Oct–Jun.* **Skin** *shiny, smooth.* **Flesh** *red-orange*

**Leaves** *dark, lustrous, yellow spots above, pale below*

This Chinese species is now widely grown commercially in the warmer regions, having been originally introduced by the European explorers, and its fruit, the familiar eating orange, often appears on the tree alongside the flowers.

# Grapefruit

## *Citrus paradisi*

**Shoot** glabrous, usually bears stout spines.
**Leaves** 15 cm, ovate, glabrous, soft, lustrous mid-green, paler below. **Petiole** to 1.5 cm across, broadly winged

**Flowers** 5 cm, white, in broad flat terminal or axillary clusters, Feb–Apr. **Leaf margins** undulate

**15 m. Crown** dense, broad, conical or rounded, sometimes reaches ground level. **Branches** pendulous at tips.
**Bole** short, often crooked.
**Fruit** always present, most ripening Nov–Apr

**Fruit** 15 cm, globose, usually single, green, ripen yellow. **Skin** very thick, smooth. **Flesh** pale, juicy

Grapefruit's large, juicy fruits have made it an important commercial tree, and it can be grown in sites unsuitable for other citrus trees as its dense crown makes it resistant to wind.

# Kumquat

## *Fortunella margarita*

**Margins** sometimes finely serrate

**Leaves** 6 cm, lanceolate or narrow elliptic. **Base** rounded

**Leaves** lustrous dark green above, paler with yellow dots below.
**Margins** slightly crinkled.
**Apex** acute

**Fruit** 3 cm, oblong or oval, pale orange, solitary or in small clusters. **Skin** spotted with translucent glands

Discovered in eastern China in the middle of the last century by the botanist Robert Fortune, Kumquat forms a small tree, sometimes only to 5 m, with a spreading crown. The small fruits are eaten whole and have a distinctive bitter-sweet flavor.

## *Euodia daniellii*

**Flowers** unisexual, in
15 cm corymbs, opening Sept

**Leaflets** (5–9) 6 cm,
cupped, very glossy,
obliquely based.
**Margins** almost
entire. **Veins** downy
below. **Midribs**
glabrous

**Buds** naked,
comprising
folded, hairy
leaves

**Shoot** and
**rachis** glabrous

**Leaves**
20–25 cm,
pinnate.
**Petiole** 3 cm

This Korean and Chinese species is valuable for the lateness of its
flowering. The Amur cork tree (*Phellodendron amurense*), noted
for its thick and coarsely ridged corky bark, has larger leaves with
7–13 leaflets and smaller, grayish buds hidden in the leaf stalk. It
is a native of northern China and Manchuria,

# Spindle family Celastraceae

## Spindle tree

### *Euonymus europaeus*

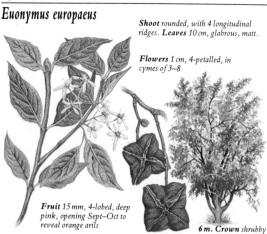

**Shoot** rounded, with 4 longitudinal
ridges. **Leaves** 10 cm, glabrous, matt.

**Flowers** 1 cm, 4-petalled, in
cymes of 3–8

**Fruit** 15 mm, 4-lobed, deep
pink, opening Sept–Oct to
reveal orange arils

**6 m. Crown** shrubby

Native to most of Europe, Spindle is a small tree or shrub
renowned for its showy fruit and the attractiveness of its purple-
red autumn foliage. Its hard, non-splintering wood was once a
useful material for making spindles and other domestic items
such as skewers, pegs and knitting needles.

# Plane family Platanaceae

While it has few botanical affiliations with the Aceraceae (pp 171–9), the leaves of this family were once mistaken for those of maples. This is reflected in the specific names of two maples — *platanoides* (p171) and *pseudoplatanus* (p172) — and in that of *Platanus acerifolia*. Planes, however, always have *alternate* leaves.

## Oriental plane

### Platanus orientalis

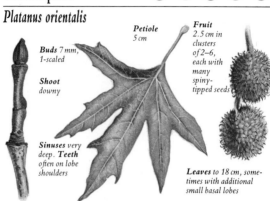

**Petiole** *5 cm*

**Fruit** *2.5 cm in clusters of 2–6, each with many spiny-tipped seeds*

**Buds** *7 mm, 1-scaled*

**Shoot** *downy*

**Sinuses** *very deep.* **Teeth** *often on lobe shoulders*

**Leaves** *to 18 cm, sometimes with additional small basal lobes*

This species from the eastern Mediterranean can grow 30 meters high and is used as a shade tree in many southern European villages. Its shortish bole can achieve a girth as large as 13 meters.

## London plane

### Platanus x acerifolia

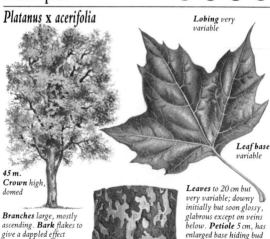

**Lobing** *very variable*

**Leaf base** *variable*

**45 m. Crown** *high, domed*

**Branches** *large, mostly ascending.* **Bark** *flakes to give a dappled effect*

**Leaves** *to 20 cm but very variable; downy initially but soon glossy, glabrous except on veins below.* **Petiole** *5 cm, has enlarged base hiding bud*

A majestic habit, an immunity to urban atmospheres and a resistance to barbaric and unsightly pollarding have countributed to this tree's popularity and its success was confirmed last century when it became widely planted in London streets. It is a hybrid of *P. orientalis* and *P. occidentalis* and has been planted across the USA from coast to coast.

## *Platanus occidentalis*

**New leaves** tomentose on both surfaces, soon becoming glabrous except along veins

**Base** of new leaves cordate, less often truncate

**Flowers** appear with new leaves, ♂ and ♀ on separate heads, single, rarely in pairs, composite

**50 m. Crown** broad, open. **Bole** straight, single, to 4 m diameter at enlarged base

**Fruit** 4 cm, rough, brown, on slender peduncle to 15 cm

**Shoot** initially densely hairy, becoming shiny orange-brown, and later gray

**Bark** at first smooth, light gray, but more characteristic where it exfoliates, revealing yellow, white or green underbark in patches

**Petiole** 12 cm

**Seed** tipped with brown hairs

**Leaves** 18 cm, broad ovate

**Leaves** thin, bright green above, paler below. **Lobes** (3–5) variable in depth, usually shallow

**Sinuses** rounded

**Bud** 12 mm, conic, lustrous, hidden in base of petiole until leaf falls

American sycamore, the tallest native broadleaf tree, is found on moist sites throughout the eastern USA. The var. *glabrata*, found in Texas· and Mexico, has less toothed leaves which are more deeply lobed, often with acute sinuses. With more deeply lobed leaves and fruits in racemes of 3–7 heads, California sycamore (*P. californica*) has dentately toothed leaves while Arizona sycamore (*P. wrightii*) has entire lobing.

# Maple family Aceraceae

Maples are noted for the rich autumnal tints of their leaves which are usually palmately lobed and always set in opposite pairs. Small green or yellow flowers appear in corymbs, racemes or panicles and are normally bisexual. Distinctive samaras (fruit keys) consist of paired seeds (nutlets) set in flat, membranous wings to facilitate wind dispersal.

## Norway maple
### Acer platanoides

**Flowers** open early Apr, before lvs in erect corymbs of 20–30. **Petals** 8 mm wide. **Buds** 1 cm tall, red-purple. **Shoot** stout, hairless, dark brown

**25 m. Crown** broad-domed, densely leaved. **Bole** short. **Bark** gray-brown; finely ridged

**Leaf** 12 × 15 cm, cordate. 2–6 coarse teeth per lobe, edges of central lobes parallel. **Blade** uns paler. **Leaf stalk** long, 15 cm, with milky sap distinguishing it from similar maples

**Samara** nutlets flat. **Wings** each 3–5 cm, almost horizontal

'Schwedleri' opens pink-red. Green by summer, then purple in autumn

'Drummondii' cv has small leaves with white or cream variegation

Native to northern and central Europe, the Norway maple is most attractive in spring when its flowers open and in autumn when its foliage turns deep yellow. This tree is suited to urban sites and a wide range of cultivars is available: 'Crimson King' has deep red-purplish leaves throughout summer.

# Sycamore maple

## *Acer pseudoplatanus*

**Shoot** green-brown. **Buds** ovoid, 1 cm, green, red margins. **Flowers** 50–100, hanging in dense, 12 cm panicles

**Samara** wings each 3 cm, set at 90°

**35 m. Crown** broad. **Branches** billow. **Bark** scaly, light gray-brown; silver, smooth on young trees. Sycamores often grown for hard, close-grained timber and to act as windbreaks

**Leaf** to 18 × 26 cm on young trees, more often 15 × 20 cm. 5 lobes, more rarely 3–7. **Teeth** coarse. **Blade** underside glaucous. **Veins** net-like. **Leaf stalk** reddish

'Erectum' cultivar (1) has fastigiate brs. Lvs of 'Purpureum' (2) are matt green with purple undersides

'Brilliantissimum' leaf (1) opens shrimp-pink, turning through yellow to dark green. 'Leopoldii' (2) can have leaves stained yellowish-pink. 'Worleei' cv leaf (3) opens in gold and yellow, fading to green-yellow

A wide tolerance of different sites and a facility to germinate, which can make it a rapacious weed, have enabled the Sycamore to colonize most of Britain. It is found across the whole of North America, although it is less common than in Britain. Its resistance to salt spray makes it ideal as a coastal tree.

# Field maple

## *Acer campestre*

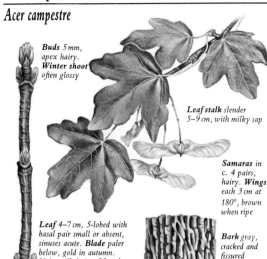

**Buds** 5 mm, apex hairy. **Winter shoot** often glossy

**Leaf stalk** slender 5–9 cm, with milky sap

**Samaras** in c. 4 pairs, hairy. **Wings** each 3 cm at 180°, brown when ripe

**Leaf** 4–7 cm, 5-lobed with basal pair small or absent, sinuses acute. **Blade** paler below, gold in autumn. **Vein axils** hairy. **Margins** ciliate

**Bark** gray, cracked and fissured

Field maple, a common hedgrow tree throughout Europe and the only maple native to Britain, makes an excellent hedge or screen plant in the more temperate regions of the USA. The Miyabe maple (*A. miyabei*), from Japan, has larger, paler leaves cut less than halfway to the leaf base.

# Italian maple

## *Acer opalus*

**Flowers** appear in April before lvs, in pendulous corymbs. **Fruit** in bunches 8–16. **Wings** each 2.5 cm, acutely angled

**Winter twig** glabrous with pale lenticels

**Leaf** 12 cm, 3- or 5-lobed. **Teeth** coarse, irregular. **Veins** impressed. **Blade** underside glaucous, hairy. **Leaf stalk** 10 cm with watery sap

Italian maple, also found in France and Spain, is a small tree reaching 15 m. Heldreich maple (*A. heldreichii*) has deep, narrow leaf sinuses—Trauvetter's maple (*A. trauvetteri*) has deep, wide ones. Both these species have *erect* flower panicles.

173

# Oregon maple

## *Acer macrophyllum*

**Shoot** *stout, green.*
**Buds** *stout, conic with red-brown scales*

**Leaf sinus** *deep, acute, rounded*

**Leaf** *paler below.* **Leaf stalk** *long, 15–35 cm, red or green, clasps stem closely, almost hiding bud.* **Samaras** *big, each wing 5 cm at 90° or less*

**Leaf blade** *very large, 20 × 35 cm, 5-lobed, thin-textured.* **Teeth** *blunt.* **Margins** *ciliate*

**Flower panicles** *narrow, 25 cm*

This west coast species, valued for its timber, grows quickly into a majestic tree of about 30 m with a tall, domed crown of ascending, arched branches. Oregon is the only maple with both milky sap *and* hanging panicles and its large leaves and clasping leaf stalks will readily confirm this identification.

# Sugar maple

## *Acer saccharum*

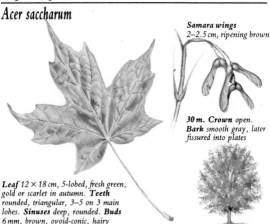

**Samara wings** *2–2.5 cm, ripening brown*

**30 m. Crown** *open.* **Bark** *smooth gray, later fissured into plates*

**Leaf** *12 × 18 cm, 5-lobed, fresh green, gold or scarlet in autumn.* **Teeth** *rounded, triangular, 3–5 on 3 main lobes.* **Sinuses** *deep, rounded.* **Buds** *6 mm, brown, ovoid-conic, hairy*

In its native range stretching from eastern Canada to Texas, this is the tree tapped for maple syrup. Its leaf is not dissimilar to that of Norway maple (p 171) but has a watery sap. Black maple (*A. Nigrum*) has darker, duller, yellow-veined leaves which are cupped and three-lobed. Its bark is ridged.

# Silver maple

## *Acer saccharinum*

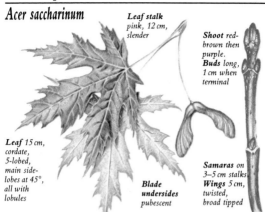

**Leaf stalk** *pink, 12 cm, slender*

**Shoot** *red-brown then purple.*
**Buds** *long, 1 cm when terminal*

**Leaf** *15 cm, cordate, 5-lobed, main side-lobes at 45°, all with lobules*

**Blade undersides** *pubescent*

**Samaras** *on 3–5 cm stalks.*
**Wings** *5 cm, twisted, broad tipped*

Silver maple is native to eastern North America south of Quebec and grows to some 30 m with a tall, raggedly domed crown and a smooth, silvery bark. In early spring its maroon flowers, opening before the leaves, enliven the tree; in autumn its foliage assumes spectacular hues of yellow, gold or red.

# Red maple

## *Acer rubrum*

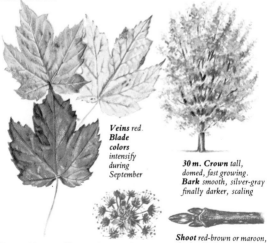

**Veins** *red.*
**Blade colors** *intensify during September*

**30 m. Crown** *tall, domed, fast growing.*
**Bark** *smooth, silver-gray finally darker, scaling*

**Leaves** *11 cm, usually dark green, glossy above, silvery below*

**Flowers** *in dense clusters, opening March, before leaves*

**Shoot** *red-brown or maroon, slender.* **Buds** *very small, 3 mm, ciliate on scales red-brown, pointed*

This North American tree is aptly named since its flowers, fruits, shoots and autumn tints are all red or reddish; its leaf undersides and bark are silver. It is closely related to Silver maple but has smaller, less deeply lobed leaves.

# Full-moon maple

## *Acer japonicum*

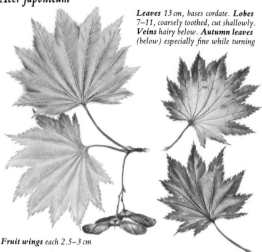

**Leaves** *13 cm, bases cordate.* **Lobes** *7–11, coarsely toothed, cut shallowly.* **Veins** *hairy below.* **Autumn leaves** *(below) especially fine while turning*

**Fruit wings** *each 2.5–3 cm*

Deriving its name from its orbicular leaves, this tree is often represented by two clones: the larger 'Vitifolium', and the smaller, yellow-leaved 'Aureum'. Vine maple (*A. circinatum*), a small tree to 10 m from the Pacific coast, has glabrous petioles.

# David maple

## *Acer davidii*

**Margins** *unevenly serrate.* **Veins** *parallel, main one red with minute axil tufts*

**Bark** *olive green with white or green "snaking" stripes.* **Branches** *ascend steeply.* **Shoot** *maroon or dark green-red, later striped white.* **Buds** *1 cm*

**Leaf blades** *ovate with pointed tips but very varied as* 'George Forrest' *(above),* 'Ernest Wilson' *(left).*

This is the most common of the snake-bark maples, which have distinctive barks with white or greenish stripes and flowers in long, dense racemes. It has a wide natural distribution in China and as a species is extremely variable. One form, 'Ernest Wilson', has a lower, rounded habit and yellower, narrower leaves folded at the base of their main vein.

# Kyushu maple

## *Acer capilles*

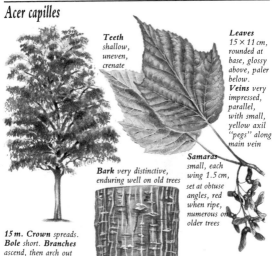

**Teeth** shallow, uneven, crenate

**Leaves** 15 × 11 cm, rounded at base, glossy above, paler below.
**Veins** very impressed, parallel, with small, yellow axil "pegs" along main vein

**Samaras** small, each wing 1.5 cm, set at obtuse angles, red when ripe, numerous on older trees

**Bark** very distinctive, enduring well on old trees

**15 m. Crown** spreads. **Bole** short. **Branches** ascend, then arch out

This Japanese maple is the only snake-bark whose leaves have small forward-pointing side lobes, impressed veins and yellowish axil "pegs". Its autumn colors and red shoots and buds give it the alternative name of "Red snake-bark maple".

# Moosewood

## *Acer pensylvanicum*

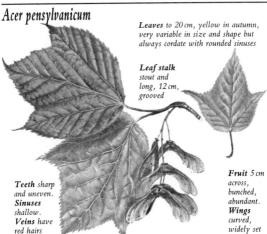

**Leaves** to 20 cm, yellow in autumn, very variable in size and shape but always cordate with rounded sinuses

**Leaf stalk** stout and long, 12 cm, grooved

**Teeth** sharp and uneven.
**Sinuses** shallow.
**Veins** have red hairs below

**Fruit** 5 cm across, bunched, abundant.
**Wings** curved, widely set

Moosewood, from eastern North America, is the only snake-bark not indigenous to eastern Asia. Honshu maple (*A. rufinerve*), from Japan, has smaller leaves with rufous, pubescent veins, buds bloomed white and small, round nutlets. Hers' maple (*A. hersii*) has matt leaves, larger fruit and no red coloring.

# Montpelier maple

## *Acer monspessulanum*

**Buds** *small, 3 mm, ovoid.* **Shoot** *slender, slightly glossy.* **Flowers** *open in June (with or before leaves) in small, erect panicles that are 5 cm wide*

**Samara** *wings small, each 1.2 cm, set almost parallel or overlapping*

**Leaf** *4 × 7 cm, cordate, usually entire.* **Blade underside** *glaucous with some axil tufts near base.* **Leaf stalk** *4 cm, slender, pink*

This maple has a wide natural range from Mediterranean Spain and north Africa to Iran and is often used for hedging, very rarely reaching its maximum height of 15 m. Cretan maple (*A. sempervirens*) is almost evergreen and has small, stiff, variable leaves, which may be unlobed or have three rounded lobes.

# Amur maple

## *Acer ginnala*

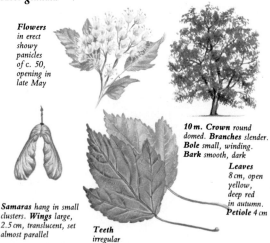

**Flowers** *in erect showy panicles of c. 50, opening in late May*

**10 m. Crown** *round domed.* **Branches** *slender.* **Bole** *small, winding.* **Bark** *smooth, dark*

**Leaves** *8 cm, open yellow, deep red in autumn.* **Petiole** *4 cm*

**Samaras** *hang in small clusters.* **Wings** *large, 2.5 cm, translucent, set almost parallel*

**Teeth** *irregular*

Amur maple, from north-east Asia, is a small shrubby tree which shows its early autumn crimson for a brief period in September. Trident maple (*A. buergeranum*) has three-lobed leaves, the lobes pointing directly forwards in the manner of Neptune's trident. These leaves have silvery undersides.

## Acer negundo

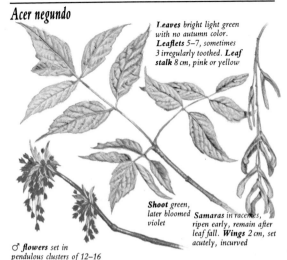

**Leaves** bright light green with no autumn color. **Leaflets** 5–7, sometimes 3 irregularly toothed. **Leaf stalk** 8 cm, pink or yellow

**Shoot** green, later bloomed violet

**Samaras** in racemes, ripen early, remain after leaf fall. **Wings** 2 cm, set acutely, incurved

♂ **flowers** set in pendulous clusters of 12–16

Possibly overplanted, Box elder rarely produces a fine specimen; its cultivar 'Variegatum' produces both fruit and leaves that are variegated white. Henry maple (*A. henryi*) is a Chinese species with red autumn foliage, pink-red leaf stalks and bright green, glossy shoots.

# Pittosporum family Pittosporaceae

## Pittosporum

### Pittosporum tenuifolium

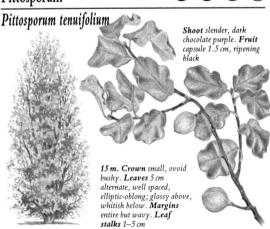

**Shoot** slender, dark chocolate purple. **Fruit** capsule 1.5 cm, ripening black

**15 m. Crown** small, ovoid bushy. **Leaves** 5 cm alternate, well spaced, elliptic-oblong; glossy above, whitish below. **Margins** entire but wavy. **Leaf stalks** 1–5 cm

The pittosporum family comes mainly from Australasia, southeast Asia and South Africa although one species is native to Madeira. Pittosporum itself is native to both islands of New Zealand and is noticeable for its fragrant flowers.

# Horse chestnut family Hippocastanaceae

This small family, named after the resemblance of its fruit to that of *Castanea* (p 120), is dominated by the *Aesculus* genus. Opposite, digitately compound leaves, showy and erect flower panicles and large seeds are identifying features.

## Horse chestnut

### Aesculus hippocastanum

**Flowers** *numerous, up to 150 on 15–30 cm panicles.* **Petals (5)** *have basal blotches*

**40 m. Crown** *tall, spreading, dense.* **Branches** *upswept, sometimes layering*

**Buds** *2.5 cm.* **Shoot** *stout.* **Leaf scars** *horseshoe shaped*

**Leaflets (5–7) 10–25 cm,** *sessile, unevenly serrate* **Fruit** *5 cm*

This species provides the ammunition for small boys to collect, throw and carve. Japanese horse chestnut (*A. turbinata*) also has sticky buds *and* sessile leaflets but the latter are evenly toothed and can be up to 40 cm long.

## Red horse chestnut

### Aesculus x carnea

**Buds** *not sticky.* **Flowers** *in 12–20 cm panicles, opening May;* 'Briottii' *flowers are deeper red*

**Bark** *similar to Horse chestnut but rougher.* **Bole** *often cankered*

This hybrid is smaller than its Horse chestnut parent while its buds resemble those of its other parent, the American Red buckeye (*A. pavia*) which has deeper red inflorescences.

# California buckeye

## *Aesculus californica*

**Leaves** palmately compound with 5 leaflets to 15 × 5 cm

**Flowers** white or pale rose, fragrant, in dense upright panicles to 25 cm in Jun–Jul. **Stamens** long

**Leaflet** oblong-lanceolate, distinctly stalked. **Margins** with fine, rounded teeth

**Apex** acuminate. **Base** rounded. **Shoot** smooth, pale green-brown. **Bud** small, acute, thickly clear-resinous

**Fruit** 10 cm, pale brown, smooth, obovoid, splits into 3 in late summer. **Seed** single, shiny orange-brown

The only buckeye native to the west, this species forms a shrubby tree to 10 m with a broad, rounded crown, and bears a poisonous seed. Indian buckeye (*A. indica*) has 7 lanceolate leaflets to 25 cm, glaucous below, and pink and yellow blotched flowers.

# Yellow buckeye

## *Aesculus flava*

**20 m. Crown** narrow with fine autumn yellows and orange-reds. **Branches** small, pendulous and twisting. **Bole** straight. **Bark** gray- or red-brown, smooth becoming scaly

**Leaves** have 5–7 leaflets with impressed veins. **Flowers** 4 cm, dense, sometimes pink, on 10–15 cm panicles. **Petals** (4) forward-pointing

**Leaflet margins** finely serrate

**Leaflet** 15 cm, glabrous, sometimes downy below, on 1.5 cm stalk. **Fruit** 6 cm, smooth, 2-seeded

Buckeyes, whose name arose when their hila (pale, basal seed scars) were likened to the eyes of deer, are native to the eastern USA. They have non-sticky buds. Ohio buckeye (*A. glabra*) has a prickly fruit and keeled bud scales. The leaflets are smaller, to 15 cm, and have an unpleasant odor when crushed.

# Linden family Tiliaceae

Lindens have large, toothed, heart-shaped leaves, flowers that hang in cymes and large distinctive bracts attached for half their length to the flower stalk. The fruit is dry and nutty.

## Large-leaved linden

### Tilia platyphyllos

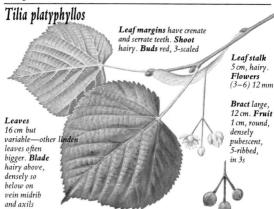

**Leaf margins** have crenate and serrate teeth. **Shoot** hairy. **Buds** red, 3-scaled

**Leaf stalk** 5 cm, hairy. **Flowers** (3–6) 12 mm

**Bract** large, 12 cm. **Fruit** 1 cm, round, densely pubescent, 5-ribbed, in 3s

**Leaves** 16 cm but variable—other linden leaves often bigger. **Blade** hairy above, densely so below on vein midrib and axils

The Large-leaved linden is native to most of Europe and has a narrow crown with branches that ascend steeply. Its bark is gray and fissured; shoots at the foot of the bole are rare. Lindens have fragrant flowers and may be planted for bee pasturage, as the honey produced from the nectar is pleasantly flavored.

## Small-leaved linden

### Tilia cordata

**Foliage** dense, slightly pendent. **Leaves** 6 cm. **Margins** finely serrate. **Blades** roundish, cordate, acuminate, shiny above, glaucous below. **Axil tufts** prominent, orange-brown

**30 m. Crown** narrow, very twiggy in winter. **Bole** can be burred

**Winter shoot** glossy brown with light lenticels. **Buds** have 2 visible scales

This species of linden is recognizable by the size of its roundish leaves and its flowers, which spread irregularly and do not hang. Mongolian linden (*T. mongolica*) is often planted where a smaller tree is required. Its leaves have large, coarse, triangular teeth which almost become lobes.

## *Tilia x europaea*

**Leaf stalk** *5 cm, hairless.* **Buds** *green or reddish.* **Flowers** *in cymes of 4–10, fragrant*

**Leaves** *10 cm, obliquely cordate.* **Blades** *glabrous below except for axil tufts, often disfigured by honey-dew left by aphids*

**40 m. Branches** *ascend and billow.* **Shoots** *sprout readily on bole and in crown.* **Bark** *ridged, heavily burred where sprouts removed*

Often the largest broadleaf in an area, the European linden is a natural hybrid of the Large- and Small-leaved species and its apparent ubiquity as a park and street tree has been attributed to seventeenth-century Dutch horticulturists who found it more easy to propagate than its parents. It is often pollarded.

## Caucasian linden

### *Tilia x euchlora*

**Shoot** *usually green but can be pink or red on outer foliage.* **Buds** *red or yellow*

**Flowers** *in cymes of 3–7.* **Bract** *long, 8 cm.* **Fruit** *hairy, 5-ribbed, tapered*

**Leaves** *10 cm but can be larger on basal sprouts.* **Blade** *obliquely cordate, glossy green with paler underside and brown axil tufts.* **Leaf stalk** *5 cm*

**20 m. Crown** *narrow, domed.* **Branches** *ascend gently then become very irregular and decurrent, thickening with age and sometimes touching the ground.* **Bole** *smooth*

Caucasian linden has uncertain origins and may be a cross between the Small-leaved linden and the rare *T. dasystyla*, also from the Caucasus. Its agreeable foliage and immunity to aphids make it more suitable for streets than European linden although its lower branches eventually become far too decurrent for such sites.

# American linden

## *Tilia americana*

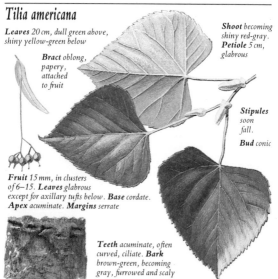

**Leaves** 20 cm, dull green above, shiny yellow-green below

**Bract** oblong, papery, attached to fruit

**Shoot** becoming shiny red-gray. **Petiole** 5 cm, glabrous

**Stipules** soon fall. **Bud** conic

**Fruit** 15 mm, in clusters of 6–15. **Leaves** glabrous except for axillary tufts below. **Base** cordate. **Apex** acuminate. **Margins** serrate

**Teeth** acuminate, often curved, ciliate. **Bark** brown-green, becoming gray, furrowed and scaly

This tall tree, native to the east from Kentucky to southern Canada, grows to 40 m, and can be distinguished from other lindens by its bark and the flowers, which grow in cymes of 6–15 and are longer than the bracts.

# White basswood

## *Tilia heterophylla*

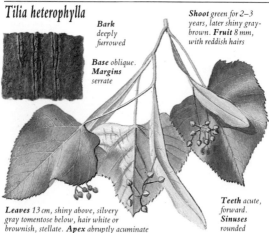

**Bark** deeply furrowed

**Base** oblique. **Margins** serrate

**Shoot** green for 2–3 years, later shiny gray-brown. **Fruit** 8 mm, with reddish hairs

**Leaves** 13 cm, shiny above, silvery gray tomentose below, hair white or brownish, stellate. **Apex** abruptly acuminate

**Teeth** acute, forward. **Sinuses** rounded

Native to inland sites in the southeastern USA, White basswood makes a tree to 25 m, distinguishable from American linden by its smaller leaves and hairy fruit. The name derives from the fibrous inner bark, or bast, once used for making ropes and brooms.

# Silver linden

## *Tilia tomentosa*

**Leaf blades** *12 × 10 cm,*
*rounded, obliquely cordate,*
*densely pubescent below.*
**Buds** *6–8 mm.* **Shoot**
*remains pubescent*

**Leaf stalk**
*to 5 cm, less*
*than half*
*blade length*

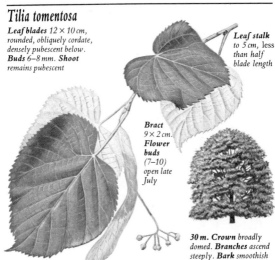

**Bract**
*9 × 2 cm.*
**Flower**
**buds**
*(7–10)*
*open late*
*July*

**30 m. Crown** *broadly*
*domed.* **Branches** *ascend*
*steeply.* **Bark** *smoothish*
*gray with cross ridges*

Silver linden displays an attractive habit, especially when its pubescent leaves are ruffled by wind. It is a native of the Balkans. Oliver linden (*T. oliveri*), from China, has larger, evenly cordate leaves and glabrous petioles and shoots.

# Pendent silver linden

## *Tilia petiolaris*

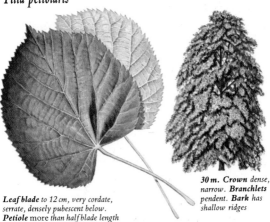

**30 m. Crown** *dense,*
*narrow.* **Branchlets**
*pendent.* **Bark** *has*
*shallow ridges*

**Leaf blade** *to 12 cm, very cordate,*
*serrate, densely pubescent below.*
**Petiole** *more than half blade length*

Usually grafted on to *T. europaea* (p 183), producing an unsightly change in bark texture at a height of some 2 m, this tree hybridizes with *T. americana* (p 184) to produce Von Moltke linden (*T.* x *moltkei*) whose larger leaves are lightly pubescent below.

# Soapberry family Sapindaceae

## Golden rain tree

### *Koelreuteria paniculata*

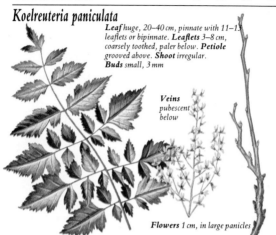

**Leaf** huge, 20–40 cm, pinnate with 11–13 leaflets or bipinnate. **Leaflets** 3–8 cm, coarsely toothed, paler below. **Petiole** grooved above. **Shoot** irregular. **Buds** small, 3 mm

**Veins** pubescent below

*Flowers 1 cm, in large panicles*

A native of Japan and China, this tree was introduced into America in 1763. It reaches a maximum height of 15 m, has a widespreading crown and is distinguished by its decurrent leaflets, showy flowers and bladder-like fruit.

# Tea family Theaceae

## Camellia

### *Camellia japonica*

**Flowers** 10 cm, with 5 spreading petals. **Flowers** solitary, sessile

**Bark** smooth, gray

**12 m. Crown** rounded or conical, often shrubby on many stems

**Leaves** leathery. **Margins** shallowly toothed

**Leaves** 10 × 5 cm, oval or ovate, shiny green above, spotted below

Commonly a garden shrub, Camellia is capable of making a small broad tree. The fruit is a brown woody capsule, and many of the 15,000 cultivars have flowers in colors other than the natural red. Sasquana camellia (*C. sasquana*) has hairy leaves and shoots.

# Japanese stewartia

## *Stewartia pseudocamellia*

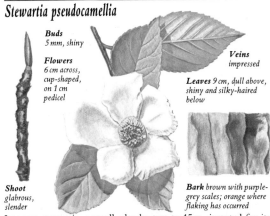

**Buds** *5 mm, shiny*

**Flowers** *6 cm across, cup-shaped, on 1 cm pedicel*

**Veins** *impressed*

**Leaves** *9 cm, dull above, shiny and silky-haired below*

**Shoot** *glabrous, slender*

**Bark** *brown with purple-grey scales; orange where flaking has occurred*

Japanese stewartia, a small, slender tree to 15 m, is noted for its attractive flowers, which remain in bloom over several weeks from July, and the brilliant colors of its autumn leaves. Mountain stewartia (*S. ovata*) has leaves with rounded bases and 10 cm flowers. The fruit is a 5-celled pointed capsule.

# Tupelo family Nyssaceae

## Tupelo

## *Nyssa sylvatica*

**Shoot** *smooth, hairless, with small lenticels.* **Buds** *5 mm, conic*

**Leaves** *5–12 cm, entire, glossy above, paler below.* **Petiole** *1 cm.* **Fruit** *1 cm, oval, in pairs on 4 cm peduncle*

*25 m.* **Crown** *conic.* **Bark** *gray, fissured*

The genus *Nyssa* comprises four species: three from eastern USA and one from China. Tupelo or Sour gum being the commonest. Water tupelo (*N. aquatica*) has larger 2.5 cm fruits and oblong-oborates leaves, often toothed, to 18 cm.

# Davidia family Davidaceae

## Dove tree  ⑥ Ⓡ 🌳 🜁 🏠

### *Davidia involucrata*

**Flower** *2 cm wide*

**20 m. Crown** *domed.* **Bole** *short.* **Branches** *radiate outwards.* **Bark** *purplish, flaking brown, with vertical fissures*

**Bracts** *large, to 17 cm, in uneven pairs, paper-thin*

**Lenticels** *pale*

**Buds** *1.5 cm, conic, shiny*

**Leaves** *15 cm, shiny above, softly hairy below.* **Veins** *impressed.* **Petiole** *15 cm.* **Vilmoriniana leaf** *(below): 20 cm, coarsely serrate, glabrous below*

**Shoot** *glabrous*

**Fruit** *3 cm, hard, fleshy, ribbed, ripening purple, with prominent lenticels.* **Pedicel** *10 cm*

Dedicated to its discoverer, Père David, this is a vigorous, linden-like tree native only to western China. Flowering is unreliable but its appearance in May can be very striking, when the large, white bracts which subtend each flower drape the branches and give the tree its name. The var. *vilmoriniana* is hardier with larger leaves.

# Myrtle family Myrtaceae

## Cider gum

### *Eucalyptus gunnii*

**Adult leaves** 10 cm, stalked, alternate

**Juvenile leaves** sessile, opposite

**Fruit** in 3s, 6 mm

**35 m. Crown** conic, with wispy halo when growing fast. **Bark** peels in orangey strips to expose smooth gray surface below

Cider gum – so called as a cider can be made from the sap – is the hardiest gum. It is a native of Tasmania and coppices readily if cut back. The flowers before opening are covered by a cap (operculum) of fused petals.

## Snow gum

### *Eucalyptus niphophila*

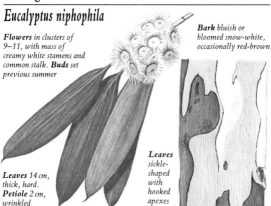

**Flowers** in clusters of 9–11, with mass of creamy white stamens and common stalk. **Buds** set previous summer

**Bark** bluish or bloomed snow-white, occasionally red-brown

**Leaves** sickle-shaped with hooked apexes

**Leaves** 14 cm, thick, hard. **Petiole** 2 cm, wrinkled

A small tree reaching 10 m, Snow gum grows wild in south-eastern Australia at heights up to 2,000 m above sea-level. Red flowering gum (*E. ficifolia*) grows to 15 m and is less hardy. Its flowers have red stamens with dark red anthers.

# Bluegum

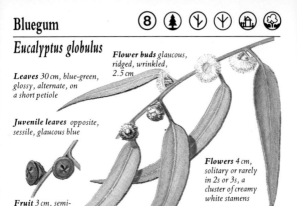

## *Eucalyptus globulus*

**Flower buds** *glaucous, ridged, wrinkled, 2.5 cm*

**Leaves** *30 cm, blue-green, glossy, alternate, on a short petiole*

**Juvenile leaves** *opposite, sessile, glaucous blue*

**Flowers** *4 cm, solitary or rarely in 2s or 3s, a cluster of creamy white stamens*

**Fruit** *3 cm, semi-sphercal, ridged, sessile, blue-white*

**Leaves** *lanceolate or sickle-shaped, with odor of camphor.* **Apex** *acuminate, often curved*

This subtropical evergreen from Australia grows to 40 m in California with a domed crown, and is recognizable by its light brown bark, which peels in long ribbons to reveal the smooth blue-gray underbark. Redgum (*E. camaldulensis*) has a white or gray bark and red flowers borne in clusters of 5–10.

# Elaeagnus family Elaeagnaceae

## Russian olive

### *Elaeagnus angustifolia*

**12 m. Crown** *rounded, spreading*

**Flowers** *1 cm, bell-shaped, short-stalked, solitary or in 3s at leaf bases.*
**Leaves** *9 cm, oblong or lanceolate*

**Bole** *often leans*

**Bark** *fissured, peeling*

**Leaves** *dull, scaly above, silvery scaled, dotted below*

**Shoot** *with silvery scales, sometimes spined*

**Fruit** *2 cm, with silvery scales.* **Flesh** *mealy*

A popular ornamental often used as a windbreak, this species from western Asia is profusely covered with silvery scales all over. It can tolerate dry climates but sometimes will not grow beyond a shrub. The fruit, containing an ellipsoid stone, is edible.

# Dogwood family Cornaceae

## Giant dogwood

### *Cornus controversa*

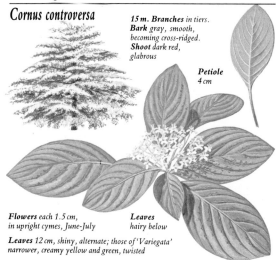

**15 m. Branches** *in tiers.*
**Bark** *gray, smooth, becoming cross-ridged.*
**Shoot** *dark red, glabrous*

**Petiole** *4 cm*

**Flowers** *each 1.5 cm, in upright cymes, June-July*

**Leaves** *hairy below*

**Leaves** *12 cm, shiny, alternate; those of 'Variegata' narrower, creamy yellow and green, twisted*

Giant dogwood is a wide-ranging tree, extending from the Himalayas to Japan. Like all dogwoods it has leaves with basal veins curving parallel to the margin but, apart from one American species, *C. alternifolia*, is unique in its alternate leaves.

## Pacific dogwood

### *Cornus nuttallii*

**Leaves** *12 cm, hairy both sides, red below in autumn.* **Petiole** *1 cm*

**Fruit** *a cluster of red drupes.*
**Flower buds** *set in autumn*

**10 m. Crown** *conic, open*

**Flowers** *green, small*

**Branches** *whorl.* **Bark** *smooth, becoming scaly*

**Flower bracts** *large, 5–8 cm, creamy, pointed, opening May (or sometimes October)*

Presenting a magnificent display when the conspicuous bracts of its flower-heads open, Pacific dogwood is a native ornamental from the west coast where it reaches 30 m. Flowering dogwood (*C. florida*) has smaller leaves and always has four bracts.

# Japanese dogwood

## *Cornus kousa*

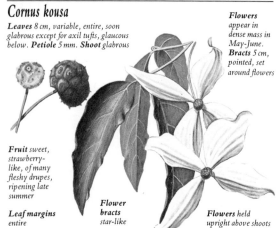

**Leaves** *8 cm, variable, entire, soon glabrous except for axil tufts, glaucous below.* **Petiole** *5 mm.* **Shoot** *glabrous*

**Flowers** *appear in dense mass in May-June.* **Bracts** *5 cm, pointed, set around flowers*

**Fruit** *sweet, strawberry-like, of many fleshy drupes, ripening late summer*

**Leaf margins** *entire*

**Flower bracts** *star-like*

**Flowers** *held upright above shoots*

This dogwood is native to Japan and central China and grows to 10 m. Its branches are somewhat tiered like those of Table dogwood, from which it can be distinguished by its opposite leaves. It is notable for its brilliant autumn reds.

# Cornelian cherry

## *Cornus mas*

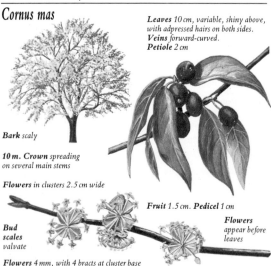

**Leaves** *10 cm, variable, shiny above, with adpressed hairs on both sides.* **Veins** *forward-curved.* **Petiole** *2 cm*

**Bark** *scaly*

**10 m. Crown** *spreading on several main stems*

**Flowers** *in clusters 2.5 cm wide*

**Fruit** *1.5 cm.* **Pedicel** *1 cm*

**Bud scales** *valvate*

**Flowers** *appear before leaves*

**Flowers** *4 mm, with 4 bracts at cluster base*

This ornamental is a native of southern Europe and has long been cultivated for the bright yellow flowers which cloak its bare branches in March and April and for its abundant fruit which provides another attractive display in late summer. This fruit is edible and can be made into jam or syrup.

# Heather family Ericaceae

## Strawberry tree

### Arbutus unedo

**Petiole** 7 mm, hairy. **Shoot** long-haired

**Leaves** 10 cm, variable, serrate, shiny, very dark above, paler below

**Flowers** 6 mm, pink or white

**Panicles** of 15–20 flowers open in Oct–Dec

**Fruit** 2 cm, warty, green-yellow, ripening to orange-pink as new flowers open

Native to south-west Ireland and southern Europe, the Strawberry tree is a small evergreen up to 10 m high and has a dark red-brown, finely fissured bark. Its strawberry-like fruit takes a year to ripen. It can be eaten but its insipidness is indicated by the tree's specific name *unedo* meaning "I eat (only) one."

## Pacific madrone

### Arbutus menziesii

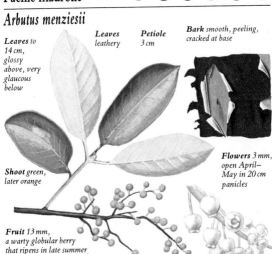

**Leaves** to 14 cm, glossy above, very glaucous below

**Leaves** leathery

**Petiole** 3 cm

**Bark** smooth, peeling, cracked at base

**Shoot** green, later orange

**Flowers** 3 mm, open April–May in 20 cm panicles

**Fruit** 13 mm, a warty globular berry that ripens in late summer.

Madrone is native to the Pacific coast where it attains 25 m and is readily identified by its bark. Similar barks are carried by the Greek strawberry tree (*A. andrachne*) which has narrower leaves and *A. x andrachnoides* with serrate leaves.

# Sorrel tree

## *Oxydendrum arboreum*

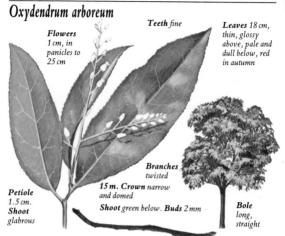

**Flowers** 1 cm, in panicles to 25 cm

**Teeth** fine

**Leaves** 18 cm, thin, glossy above, pale and dull below, red in autumn

**Branches** twisted

**15 m. Crown** narrow and domed

**Shoot** green below. **Buds** 2 mm

**Petiole** 1.5 cm. **Shoot** glabrous

**Bole** long, straight

Native to the eastern USA, Sorrel tree of Sourwood flowers into autumn, with its foliage often already turning scarlet. Its name refers to the pleasantly sour taste of its leaves. The fruit capsules are whitish and 12 mm wide on long peduncles, and split into 5 segments.

# Ebony family Ebenaceae

## Persimmon

### *Diospyros virginiana*

**Flowers** 2 cm, open in July

**Leaves** dark glossy above

**Bark** black or dark gray, cracked into squares

**Fruit** 4 cm, orange or purple. **Calyx** 4-lobed, persistent

**Leaves** to 15 cm, waved, paler below

**Shoot** slender, downy or glabrous. **Buds** small

Persimmon, from the south-eastern USA, grows up to 20 m and is distinguished by its glossy leaves and astringent fruit which is edible after exposure to autumn frosts. Date plum (*D. lotus*), an Asian relative, has petioles only 1 cm long and 8 mm white flowers. Kaki (*D. kaki*) has young shoots with brownish hairs.

194

# Storax family Styracaceae

## Japanese snowbell

### *Styrax japonica*

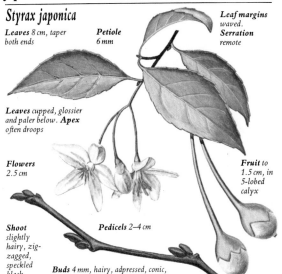

**Leaves** 8 cm, taper both ends

**Petiole** 6 mm

**Leaf margins** waved. **Serration** remote

**Leaves** cupped, glossier and paler below. **Apex** often droops

**Flowers** 2.5 cm

**Fruit** to 1.5 cm, in 5-lobed calyx

**Shoot** slightly hairy, zig-zagged, speckled black

**Pedicels** 2–4 cm

**Buds** 4 mm, hairy, adpressed, conic, pale greenish-brown

A native of Japan and China, Snowbell tree grows to 10 m with a dense, rounded crown of horizontal branches. Storax (*S. Officinalis*) is a smaller tree with white down on the young twigs, leaves and flowers. It has ovate, cordate leaves and round fruits.

## Fragrant snowbell

### *Styrax obassia*

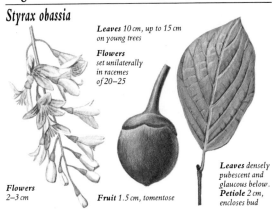

**Leaves** 10 cm, up to 15 cm on young trees

**Flowers** set unilaterally in racemes of 20–25

**Leaves** densely pubescent and glaucous below. **Petiole** 2 cm, encloses bud

**Flowers** 2–3 cm

**Fruit** 1.5 cm, tomentose

Native to Japan, Fragrant snowbell has a gray bark and an open, upright crown that reaches 15 m. Hemsley snowbell (*S. hemsleyana*) carries its white flowers on short, pubescent racemes. The petioles of its less downy leaves do not enclose the bud.

195

## *Halesia monticola*

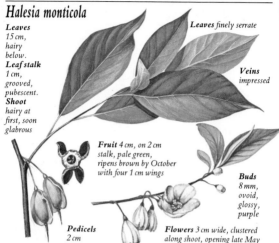

**Leaves**
15 cm,
hairy
below.
**Leaf stalk**
1 cm,
grooved,
pubescent.
**Shoot**
hairy at
first, soon
glabrous

**Leaves** *finely serrate*

**Veins**
*impressed*

**Fruit** *4 cm, on 2 cm
stalk, pale green,
ripens brown by October
with four 1 cm wings*

**Buds**
*8 mm,
ovoid,
glossy,
purple*

**Pedicels**
*2 cm*

**Flowers** *3 cm wide, clustered
along shoot, opening late May*

Mountain silverbell, to 25 m, is native to the mountains of the south-eastern USA and is distinguished by its fruit and flowers. Carolina silverbell (*H. carolina*) is shrubbier with smaller flowers.

# Loosestrife family Lythraceae

## *Lagerstroemia indica*

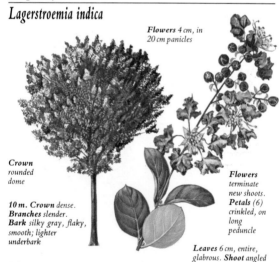

**Flowers** *4 cm, in
20 cm panicles*

**Crown**
*rounded
dome*

**10 m. Crown** *dense.*
**Branches** *slender.*
**Bark** *silky gray, flaky,
smooth; lighter
underbark*

**Flowers**
*terminate
new shoots.*
**Petals** *(6)
crinkled, on
long
peduncle*

**Leaves** *6 cm, entire,
glabrous.* **Shoot** *angled*

This native of China and Japan only produces its splendid flowers in climates with long, hot summers. The flowers are usually bright pink but can be white, purple or scarlet. The buds and leaves may be set in pairs, threes or singly along the same shoot.

# Olive family Oleaceae

The main features of this family of about 400 trees and shrubs are the opposite, simple or pinnate leaves and the perfect or unisexual flowers with 2 stamens. The fruit may be a drupe, capsule or samara. Ash is the main tree genus and has fissured barks and shoots flattened between the buds. The fruit is a samara while *Olea*, *Ligustrum* and *Phillyrea* have simple leaves and a drupaceous fruit.

## European ash

### *Fraxinus excelsior*

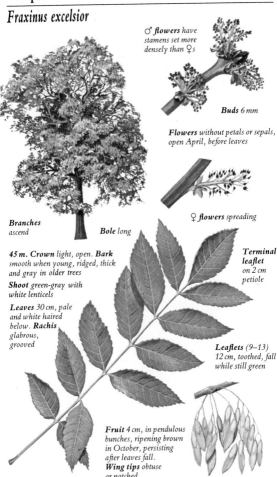

*♂ flowers* have stamens set more densely than ♀s

**Buds** 6 mm

**Flowers** without petals or sepals, open April, before leaves

**Branches** ascend

**Bole** long

♀ *flowers* spreading

**45 m. Crown** light, open. **Bark** smooth when young, ridged, thick and gray in older trees
**Shoot** green-gray with white lenticels

**Leaves** 30 cm, pale and white haired below. **Rachis** glabrous, grooved

**Terminal leaflet** on 2 cm petiole

**Leaflets** (9–13) 12 cm, toothed, fall while still green

**Fruit** 4 cm, in pendulous bunches, ripening brown in October, persisting after leaves fall.
**Wing tips** obtuse or notched

Native throughout Europe and Asia Minor, this ash is a large forest tree that grows best on heavy, alkaline loams. While male and female flowers normally appear on separate trees, one tree may carry both sexes. The squat, black buds and smooth twigs are its best identification features. 'Pendula' is a weeping form, while 'Diversifolia' has simple leaves.

## Fraxinus americana

**40 m. Crown** open, conical. **Branches** stout, spreading, pendent in lower crown, often very angular.
**Bole** often forked, sometimes repeatedly

**Bark** gray, smooth at first but becoming evenly fissured into diamond-shaped ridges; older trees deeply fissured

**Flowers** open before leaves, in glabrous panicles, dioecious, without petals, looking like tassels; ♂ and ♀ on separate trees

**Leaves** 30 cm. **Leaflets** (5–9) 12 cm, ovate or oblong-lanceolate. **Petiolule** 1 cm, except on terminal leaflet, to 4 cm

**Shoot** stout, soon glabrous, often bloomed, gray or brown

**Bud** obtuse or rounded, broadly ovate. **Scales** in 4 pairs, outer 2 keeled, acute, hairy

**Fruit** to 6 cm × 6 mm

**Base** cuneate. **Apex** acute

**Fruit** a samara, in pendent clusters to 20 cm. **Wing** pointed or notched at tip

**Samara** ripens from bright green to dull brown

**Leaflets** gray-green below. **Veins** prominent; those on underside may be pubescent. **Margins** either entire or serrate, with forward teeth

**Leaves** turn dull gray in autumn, rarely yellow or purple. **Samara** wing decurrent

American or white ash is found throughout the east from Nova Scotia to Texas growing on moist soils, often near streams, and in the open, not being tolerant of much shade. This is the commonest ash, and is distinguishable from others by its larger leaves and black winter buds. Oregon ash (*F. latifolia*) has 35 cm leaves with 5 or 7 elliptic or ovate 17 cm leaflets, sessile or nearly so, which have entire or finely toothed margins. The conic buds and shoots are hairy.

# Green ash

## *Fraxinus pennsylvanica*

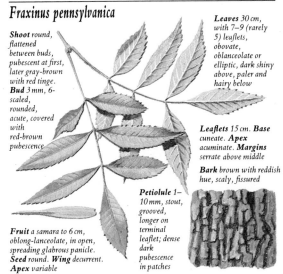

**Shoot** round, flattened between buds, pubescent at first, later gray-brown with red tinge.
**Bud** 3 mm, 6-scaled, rounded, acute, covered with red-brown pubescence

**Leaves** 30 cm, with 7–9 (rarely 5) leaflets, obovate, oblanceolate or elliptic, dark shiny above, paler and hairy below

**Leaflets** 15 cm. **Base** cuneate. **Apex** acuminate. **Margins** serrate above middle

**Bark** brown with reddish hue, scaly, fissured

**Petiolule** 1–10 mm, stout, grooved, longer on terminal leaflet; dense dark pubescence in patches

**Fruit** a samara to 6 cm, oblong-lanceolate, in open, spreading glabrous panicle. **Seed** round. **Wing** decurrent. **Apex** variable

With a wide natural distribution throughout eastern and central North America, Green or Red ash grows to 20 m and is variable in habit, the pubescence on leaf and shoot, the leaflet shape, and the rounded, acute or notched samara wing.

# Black ash

## *Fraxinus nigra*

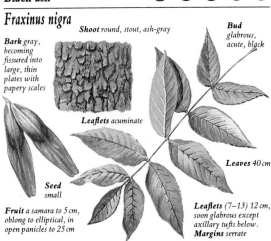

**Shoot** round, stout, ash-gray

**Bud** glabrous, acute, black

**Bark** gray, becoming fissured into large, thin plates with papery scales

**Leaflets** acuminate

**Leaves** 40 cm

**Seed** small

**Fruit** a samara to 5 cm, oblong to elliptical, in open panicles to 25 cm

**Leaflets** (7–13) 12 cm, soon glabrous except axillary tufts below. **Margins** serrate

Black ash closely resembles Blue ash (p200), its flowers having no calyxes, but it differs in its round shoots and sessile leaflets. It is native to the northeast, preferring wet sites, and reaches 25 m with an open, narrow crown.

199

# Flowering ash

## *Fraxinus ornus*

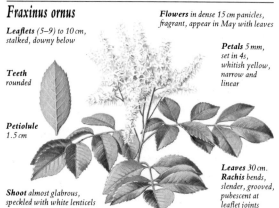

**Flowers** in dense 15 cm panicles, fragrant, appear in May with leaves

**Leaflets** (5–9) to 10 cm, stalked, downy below

**Petals** 5 mm, set in 4s, whitish yellow, narrow and linear

**Teeth** rounded

**Petiolule** 1.5 cm

**Leaves** 30 cm. **Rachis** bends, slender, grooved, pubescent at leaflet joints

**Shoot** almost glabrous, speckled with white lenticels

Flowering ash reaches 20 m to form a tree similar to *F. excelsior*, but distinguished from it, when not in flower, by its stalked, pubescent leaflets and gray-brown, hairy buds. It grows wild in southern Europe and western Asia and is sometimes tapped for the sugary substance which exudes from the bark.

# Blue ash

## *Fraxinus quadrangulata*

**Leaflets** (5–11) paler below

**Bud** 6 mm, rounded, red-brown, slightly hairy, with 6 scales. **Leaflets** serrate. **Apex** acuminate. **Base** oblique

**Shoots** gray, 4-angled, with 4 narrow corky wings, becoming round in 3rd season

**Fruit** a samara, 5 cm, oblong-ovate, loosely clustered in panicles. **Seed** flattened. **Wing** thin, decurrent, rounded, acute or notched at apex, ribbed

**Leaflet base** oblique

**Leaves** 30 cm. **Rachis** slender

Blue ash, native to a restricted area south of the Great Lakes, is usually a small tree to 15 m with a slender crown, although it occasionally reaches 40 m. The bark is irregularly divided into large plates, the light gray surface being tinged with red. It takes its name from a dye made from the inner bark.

# Caucasian ash

## *Fraxinus oxycarpa*

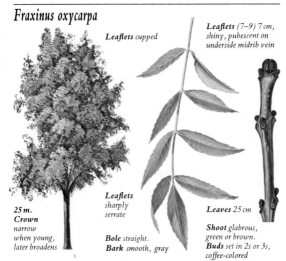

**Leaflets** cupped

**Leaflets** (7–9) 7 cm, shiny, pubescent on underside midrib vein

**Leaflets** sharply serrate

**Leaves** 25 cm

**25 m. Crown** narrow when young, later broadens

**Bole** straight. **Bark** smooth, gray

**Shoot** glabrous, green or brown. **Buds** set in 2s or 3s, coffee-colored

This ash grows wild from southern Europe across the Caucasus to Iran and is usually encountered as the clone 'Raywood', whose leaves turn claret in autumn. Narrow-leafed ash (*F. angustifolia*) has slender, glabrous leaflets and a rougher, dark gray bark.

# Velvet ash

## *Fraxinus velutina*

**Buds** 5 mm, velvety, 6-scaled

**Fruit** 2 cm on downy pedicel. **Wing** shorter than seed, notched

**Leaflets** downy

**Shoot** round, slender, velvety in 1st year

**Leaflets** 5 cm, thick, bluntly toothed above middle, usually in 5s or, less often, 3s, 7s or 9s

**Leaf** 15 cm

Although some forms are almost glabrous, Velvet or Arizona ash, from the south-western USA and Mexico, has pubescence even on its flower panicles. It can stand great extremes of temperature and reaches 10 or 15 m. The bark is broadly ridged.

# Olive

## *Olea europaea*

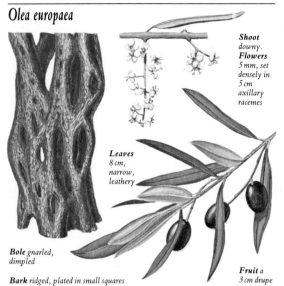

**Shoot** downy.
**Flowers** 5 mm, set densely in 5 cm axillary racemes

**Leaves** 8 cm, narrow, leathery

**Bole** gnarled, dimpled

**Bark** ridged, plated in small squares

**Fruit** a 3 cm drupe

A common feature of the Californian and Mediterranean landscapes, the Olive has a densely branched crown that reaches 15 m as an orchard tree but in the wild is much shrubbier and has small, oval leaves. If to be eaten, the fruit is harvested when green; for oil production it is left until black and fully ripe.

# Glossy privet

## *Ligustrum lucidum*

**Flowers** creamy-white, fragrant, in many 15 cm conic panicles

**Leaves** 10 cm, entire, folded along midrib, dark, glossy above, paler and mat below

**15 m. Crown** dense

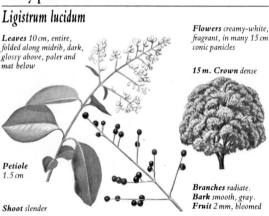

**Petiole** 1.5 cm

**Branches** radiate.
**Bark** smooth, gray.
**Fruit** 2 mm, bloomed

**Shoot** slender

This superb Chinese evergreen is noted for its glossy and leathery leaves and the lateness of its flowers which open in late summer. Lilac (*Syringa vulgaris*) flowers in May and June and is deciduous with stout shoots ending in large pairs of green buds.

# Swamp privet

## *Forestiera acuminata*

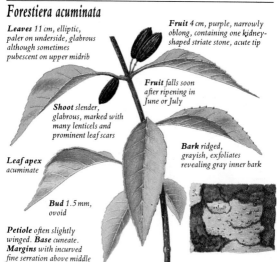

**Leaves** 11 cm, elliptic, paler on underside, glabrous although sometimes pubescent on upper midrib

**Fruit** 4 cm, purple, narrowly oblong, containing one kidney-shaped striate stone, acute tip

**Fruit** falls soon after ripening in June or July

**Shoot** slender, glabrous, marked with many lenticels and prominent leaf scars

**Bark** ridged, grayish, exfoliates revealing gray inner bark

**Leaf apex** acuminate

**Bud** 1.5 mm, ovoid

**Petiole** often slightly winged. **Base** cuneate. **Margins** with incurved fine serration above middle

Swamp privet is found in wet sites in the lower Mississippi region and as far east as the Georgia coast, and forms a small tree to 15 m with a spreading crown. The flowers, which have no petals, appear on short pedicels in many flowered fasicles.

---

# Fringe tree

## *Chionanthus virginicus*

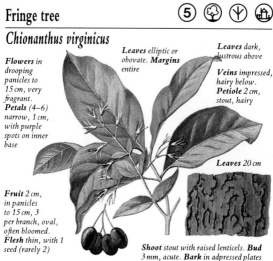

**Flowers** in drooping panicles to 15 cm, very fragrant. **Petals** (4–6) narrow, 1 cm, with purple spots on inner base

**Leaves** elliptic or obovate. **Margins** entire

**Leaves** dark, lustrous above

**Veins** impressed, hairy below. **Petiole** 2 cm, stout, hairy

**Leaves** 20 cm

**Fruit** 2 cm, in panicles to 15 cm, 3 per branch, oval, often bloomed. **Flesh** thin, with 1 seed (rarely 2)

**Shoot** stout with raised lenticels. **Bud** 3 mm, acute. **Bark** in adpressed plates

This southeastern species grows to 10 m with a narrow, oblong crown, and is widely planted as an ornamental, being valued for its attractive star-like flowers which appear in June on the previous year's shoots below the expanding leaves.

# Phillyrea

## *Phillyrea latifolia*

**Leaves** 6 cm, variable, opposite, with 7–12 pairs of lateral veins

**Fruit stalk** 1 cm

**Surface** glossy above, mat beneath with prominent veins

**Fruit** 1 cm, ripens red to blackish purple

**Midrib** raised

**Apex** acute. **Buds** minute

**Margins** toothed or entire. **Shoot** initially downy

Phillyrea, found in evergreen woods in the Mediterranean region, reaches 10 m. It has a rounded crown with very glossy foliage which appears almost black, and a dense, shrub-like habit. The bark is smooth and gray. Its small, whitish green flowers appear in short axillary clusters during June.

# Figwort family Scrophulariaceae

## Paulownia

### *Paulownia tomentosa*

**Margins** entire or with 4–6 teeth

**Leaves** 35 cm, texture very thin

**Flowers** 5 cm in panicles to 30 cm, fragrant

**Shoot** stout with prominent lenticels, initially downy

**Petiole** 15 cm, downy, pink-yellow

**Buds** minute

**Calyx** downy, pale brown, covers winter flower buds

**Flowers** bell-like

A native of China, Paulownia, also called Empress tree, has a gaunt, domed crown and reaches 20 m. The flowers are often damaged over winter when fully exposed in bud. They are followed by green, ovoid, pointed capsules containing winged seeds.

# Bignonia family Bignoniaceae

## Indian bean tree ● Northern catalpa

### *Catalpa bignonioides ● Catalpa speciosa*

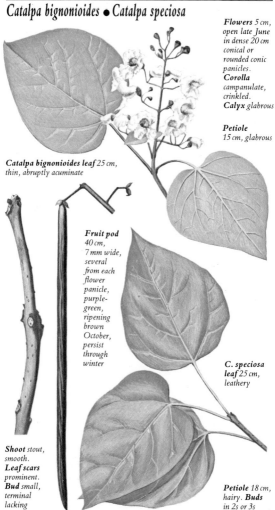

**Flowers** *5 cm,
open late June
in dense 20 cm
conical or
rounded conic
panicles.*
**Corolla**
*campanulate,
crinkled.*
**Calyx** *glabrous*

**Petiole**
*15 cm, glabrous*

*Catalpa bignonioides leaf 25 cm,
thin, abruptly acuminate*

**Fruit pod**
*40 cm,
7 mm wide,
several
from each
flower
panicle,
purple-
green,
ripening
brown
October,
persist
through
winter*

**C. speciosa
leaf** *25 cm,
leathery*

**Shoot** *stout,
smooth.*
**Leaf scars**
*prominent.*
**Bud** *small,
terminal
lacking*

**Petiole** *18 m,
hairy.* **Buds**
*in 2s or 3s*

The catalpas are a small group of trees with large ovate leaves and
long, hanging pods. The commonest, Indian bean tree, from the
Gulf Coast, now widespread, forms a low, spreading tree,
occasionally to 20 m. Northern catalpa, from central USA, has a
conic crown and may attain 40 m, or half this height in Europe.
Its flowers, to 6 cm, open earlier and the calyx is hairy. The
Chinese Yellow catalpa (*C. ovata*) has yellower 3 cm flowers and
broadly ovate 3-lobed leaves with long tips. Hybrid catalpa (*C. x
erubescens*) has leaves opening purple.

# Desert willow

## *Chilopsis linearis*

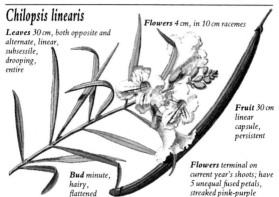

**Flowers** 4 cm, in 10 cm racemes

**Leaves** 30 cm, both opposite and alternate, linear, subsessile, drooping, entire

**Fruit** 30 cm linear capsule, persistent

**Bud** minute, hairy, flattened

**Flowers** terminal on current year's shoots; have 5 unequal fused petals, streaked pink-purple

This large bush or small tree from the southwest, which grows to only 8 m, is similar in fruit to Catalpa, but can be distinguished by the linear leaves, often erratically arranged on the shoot. The flowers have only two, not four, stamens.

# Palm family Palmaceae

## Chusan palm

## *Trachycarpus fortunei*

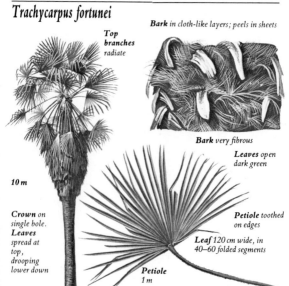

**Top branches** radiate

**Bark** in cloth-like layers; peels in sheets

**Bark** very fibrous

**Leaves** open dark green

**10 m**

**Petiole** toothed on edges

**Crown** on single bole.
**Leaves** spread at top, drooping lower down

**Leaf** 120 cm wide, in 40–60 folded segments

**Petiole** 1 m

Chusan palm, also called Windmill palm, from central China, is the palm most tolerant of cold and remarkable for its cloth-like bark. Like all true palms it has only a single growing point and so never branches.

# Royal palm

## *Roystonea regia*

**30 m. Crown** *an apical cluster of foliage.*
**Bole** *long, often swollen at middle*

**Bark**
blotchy
in color,
broken into
small,
irregular
plates.
**Leaf rachis**
hemispherical,
unarmed, scaly,
clasping stem
at base.
**Leaves** 3–
6 m, closely
pinnate,
arched

**Fruit** *a 1.3 cm drupe,
blue-black, subsessile,
with thin loose outer coat*

**Leaflets**
1 m, up to
3 cm wide
at base but
tapering
toward
apex,
numerous,
with
glandular
spots below
and many
parallel
veins

Native from southern Florida throughout central America, this palm is widely planted in humid subtropical regions as an ornamental, often being used in avenues of trees. The bole, often swollen halfway up, is diagnostic, and the small, white flowers are clustered in spadices (spikes with fleshy axes).

# Date palm

## *Phoenix dactylifera*

**Young leaves** *erect, older leaves
arched and pendent.* **Leaves**
6 m, pinnate, with many
pairs of feathery
leaflets. **Leaflets**
50 cm, gray-green,
glaucous below,
often drooping.
**Apex** long,
acute

**Seed**
deeply
grooved

**Fruit**
4 cm,
a dark,
lustrous
oblong
berry, in
dense
clusters
on ♀
trees.
**Flesh**
edible,
sugary

**30 m.
Crown**
unbranched.
**Bole**
covered
with dead
leaf bases,
may put out
suckers at
base

**Leaflet** *folded at base*

This fruit tree is widely cultivated in the hot, dry regions of Florida and California which resemble its native North Africa and western Asia. Small, yellowish flowers are produced in panicles but, being dioecious, artificial pollination is used.

# Canary Island Date palm

## *Phoenix canariensis*

**Leaves** to 7 m, erect and arching, pinnate, evergreen.
**Leaflets** 50 cm, narrow, straight, folded at base with margins turned upward. **Apex** acuminate, with very long, tapered point

**20 m. Crown** of broad, arching leaves

**Fruit** yellow, with long 1 m peduncle

**Rachis** very stout, with two long green thorns at base of leaflets.
**Leaf** persists hanging on tree when dead

**Fruit** sessile

**Fruit** 2 cm, an ovoid or rounded drupe

**Flesh** dry, inedible

**Bole** very stout, often widening near top and at base where aerial roots may protrude; rough, covered with pale brown leaf bases

Sometimes called Pineapple palm because of the appearance of its bole, this species can be distinguished from the similar Date palm (p207) by its thicker stem, wide, stiffer leaflets and the dry, inedible fruit. It is named after its native islands.

# European fan palm

## *Chamaerops humilis*

**Leaves** fan-shaped, to 50 cm

**Leaves** corrugated, split nearly to base into 12-15 folded, acute segments

**Petiole** 1 m, slender, flat above, rounded below, with forward pointing spines.
**Flowers** emerge from spathe in cluster

**Bark** not visible.
**Bole** covered with upward-pointing hemispherical bases of dead leaves; near top also bears stiff, dark fibers

In its native habitat on the shores of the western Mediterranean this palm may occur as a dwarf shrub or a small tree to 6 m, but in cultivation in subtropical North America it can grow to 10 m. The small yellow flowers, in clusters, are followed by brown or yellow rounded fruits to 4 cm.

# California fan palm

## 8

## *Washingtonia filifera*

**Margins** *frayed, with long threads*

**20m. Crown** *spreading, arching above, with shaggy skirt of drooping dead leaves around bole*

**Bark** *smooth, reddish brown, with scars from fallen leaves*

**Pulp** *dry.* **Seed** *(1) brown*

**Fruit** *1 cm, an oval black berry, in branched 3 m clusters*

**Leaves** *1.5 m, palmate, pale green, made up of numerous narrow, folded, leathery segments.* **Petiole** *1.5 m, hooked*

Widely planted as a street ornamental, this is one of the largest palms, and is distinctive for the way its leaves arise erect and end up as a dead frill around the trunk. It differs from many palms in its smooth bark without fibrous threads.

# Mexican fan palm

## 9

## *Washingtonia robusta*

**20 m. Crown** *narrow*

**Leaves** *1.5 m, palmate, bright green, divided into many narrow, folded leathery segments*

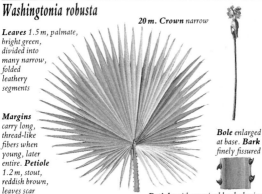

**Margins** *carry long, thread-like fibers when young, later entire.* **Petiole** *1.2 m, stout, reddish brown, leaves scar*

**Bole** *enlarged at base.* **Bark** *finely fissured*

**Petiole** *with marginal hooked spines*

Mexican fan palm is similar to California fan palm, but has a narrower crown, less smooth bark and loses the marginal fibers as the leaves mature. It is also native to California, now widely planted across the Gulf states, but prefers temperate coastal sites.

# Florida thatch palm

## *Thrinax parviflora*

**10 m. Crown**
a rounded apical
cluster of foliage.
**Petiole** 1.5 m, green,
reddish near stem, broad,
flattened, with winged margin

**Leaves** 1 m across, to 60 cm long, fan-shaped, divided
nearly to the center into many linear, folded segments
with long-pointed apexes

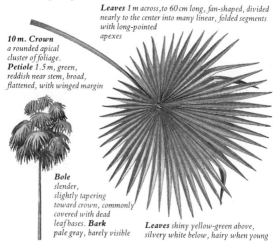

**Bole**
slender,
slightly tapering
toward crown, commonly
covered with dead
leaf bases. **Bark**
pale gray, barely visible

**Leaves** shiny yellow-green above,
silvery white below, hairy when young

The thatch palms are a group of small trees found mainly in the Caribbean. Florida thatch palm, one of four native to the south of that state, is the commonest, and is also known as Jamaica thatch palm. The fragrant yellow flowers appear in a raceme.

# Cabbage palmetto

## *Sabal palmetto*

**Bark** at first with
basal sheaths
of dead
leaves

**Leaves** 1.5 m long by
2 m wide, palmate
**Petiole** 2 m,
unarmed,
hemi-spherical

**Margins**
fibrous

**Bark** develops vertical
fissures

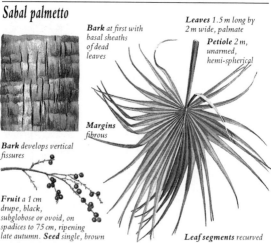

**Fruit** a 1 cm
drupe, black,
subglobose or ovoid, on
spadices to 75 cm, ripening
late autumn. **Seed** single, brown

**Leaf segments** recurved

Native to Florida and up the coast to South Carolina, Cabbage palmetto grows to 25 m with a thatch of dead leaves below its crown like California fan palm, though differing in its unarmed petioles. The bark becomes smooth with age.

# Agave family Agavaceae

## Cabbage palm

### *Cordyline australis*

**Leaves** 90 cm, lanceolate, pointed tip, erect, later drooping, persisting dead

*10 m*

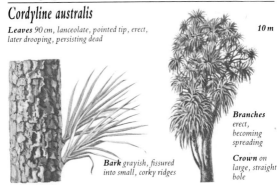

**Branches** erect, becoming spreading

**Crown** on large, straight bole

**Bark** grayish, fissured into small, corky ridges

Cabbage palm, from New Zealand, is hardy on warm coasts. Leaves are arranged spirally in tufts on shoots. The fragrant, creamy-white flowers are carried in terminal panicles up to 1 m high and are followed by numerous 6 mm bluish-white berries.

## Joshua tree

### *Yucca brevifolia*

**10 m.**
**Crown** erect, growing broader with height and forking at every old flower point

**Leaves** 40 cm, dagger-like with broad base, ending in sharp spine, glaucous. **Margins** yellowish, sharply toothed. **Teeth** very small

**Leaves** clustered at ends of stout, sparse branches. **Fruit** 10 cm, 3-angled, ripens brown

**Flowers** 5 cm, in pubescent panicles Mar–May

**Flowers** open from conic 25 cm bud. **Petals** waxy, incurved, keeled, slightly united at base

**Fruit** in spreading or pendent clusters. **Bark** rough, corky, deeply furrowed into 60 cm plates

This southwestern species, a native of the Mohave Desert, was named by the early Mormons who thought its habit resembled a man reaching out his arms to heaven. It is the tallest of several Yuccas from the south although it has the shortest leaves.

# Index

Entries refer to spp, vars and cvs described in detail and illustrated; for related trees see genus or nearest sp entry.